TECHNOLOGIES
IN THE ERA OF
SINGULARITY

GURU PADA CHATTOPADHYAY

INDIA • SINGAPORE • MALAYSIA

Notion Press

Old No. 38, New No. 6
McNichols Road, Chetpet
Chennai - 600 031

First Published by Notion Press 2018
Copyright © Guru Pada Chattopadhyay 2018
All Rights Reserved.

ISBN 978-1-64324-145-6

This book has been published with all reasonable efforts taken to make the material error-free after the consent of the author. No part of this book shall be used, reproduced in any manner whatsoever without written permission from the author, except in the case of brief quotations embodied in critical articles and reviews.

The Author of this book is solely responsible and liable for its content including but not limited to the views, representations, descriptions, statements, information, opinions and references ["Content"]. The Content of this book shall not constitute or be construed or deemed to reflect the opinion or expression of the Publisher or Editor. Neither the Publisher nor Editor endorse or approve the Content of this book or guarantee the reliability, accuracy or completeness of the Content published herein and do not make any representations or warranties of any kind, express or implied, including but not limited to the implied warranties of merchantability, fitness for a particular purpose. The Publisher and Editor shall not be liable whatsoever for any errors, omissions, whether such errors or omissions result from negligence, accident, or any other cause or claims for loss or damages of any kind, including without limitation, indirect or consequential loss or damage arising out of use, inability to use, or about the reliability, accuracy or sufficiency of the information contained in this book.

Dedication

I dedicate this book to the two young ladies. Mala, my wife, for patiently bearing with me for over a year while I merrily punched away on my Mac. And Sagarika, my daughter, for taking pains in editing the manuscript and giving me useful tips in shaping the book.

CONTENTS

Preface .... *ix*

CHAPTER 1: SINGULARITY 1
1.1 What is Singularity 3
1.2 Technological Singularity 4
1.3 Consequences of Singularity 4
1.4 Will Singularity be Real 5
1.5 Technologies Leading to Singularity 8
1.6 Why Are We Skeptical about Exponential Future? 10
1.7 Ethics in Singularity 12
References 13

CHAPTER 2: LIBERAL ARTS EDUCATION 15
2.1 Why Liberal Arts 15
2.2 Quotes from World Economic Forum 15
2.3 Skills of 21st Century 17
2.4 How will Liberal Education Help 19
2.5 Liberal Arts in India 20
References 25

CHAPTER 3: ROBOTICS AND ARTIFICIAL INTELLIGENCE ... 27
Robot 27
3.1 Laws of Robotics 28
3.2 Introduction 28
3.3 Definitions 30
3.4 Robot Revolution 32
3.5 Robot Structure 32
3.6 Artificial Intelligence 35
3.7 AI Today 40

3.8 Machine Learning 45
3.9 Natural Language Processing 47
3.10 Speech Recognition 50
3.11 Computer Vision 53
3.12 Future Impact of AI in Developed Cities 56
3.13 Will AI be Conscious 59
3.14 AI & Big Corporates 61
3.15 Robotics & AI Courses 62
3.16 Start-ups in India 65
3.17 Robot Olympiad 68
3.18 Conclusion 70
References 71
Appendix 72

CHAPTER 4: NANOTECHNOLOGY 77

Introduction 77
4.1 Definitions 80
4.2 What is Special about Nano 81
4.3 Nanoscience and Nanotechnology in Nature 83
4.4 Nanomanufacturing 87
4.5 Type of Nanomaterials 94
4.6 Imaging of Nanomaterials 99
4.7 Applications of Nanotechnology 102
4.8 Two Concepts 113
4.9 Nanotechnology Initiative of Govt. of India 114
4.10 Nanotechnology Companies in India 117
4.11 Nanotechnology Courses 123
4.12 Nanotechnology in Schools 129
4.13 Risk of Nanotechnology 130
4.14 Ethical, Legal and Societal Issues 132
4.15 Conclusion 133
References 135

CHAPTER 5: GENETICS 137

Introduction 137
5.1 Definitions: At the very outset, let's define few terms 143
5.2 Genetics, Gene and DNA 145
5.3 Function of Genes 148

5.4 DNA and RNA	150
5.5 Genetic Variation	150
5.6 Genes and Environment	151
5.7 Genomics	153
5.8 Population Genetics	155
5.9 Quantitative Genetics	155
5.10 DNA Sequencing	156
5.11 Whole Genome Sequencing	157
5.12 Importance of Genetics and Genomics in Health	157
5.13 Human Genome Project	159
5.14 Technology and Techniques	160
5.15 Bioinformatics	162
5.16 Genetic Engineering	165
5.17 Synthetic Biology	169
5.18 Epigenetics	173
5.19 Personalised Medicine and Digital Patient	174
5.20 Internet of Things in Health	180
5.21 Few Genetics Courses, Societies, Journals	188
5.22 Conclusion	190
References	191
Epilogue	*193*

PREFACE

We normally gaze at the crystal ball to extrapolate the future from the present, a practice which has worked well so far. But times are changing. We now live in one of the most exciting times of history and have reached at the inflection point of an exponential curve in the field of science and technology. From mid 2014 to 2025, there will be seven 18 months "Moore's Law" generations resulting in a 128-fold increase in computing performance. That will enable devices with core components to be more powerful, five times cheaper and five times smaller. In the same period, there will be striking development in areas of artificial intelligence, deep learning, analytics, big data, genetics and nanotechnology. Concurrently with this exponential growth, these technologies will not be standalone pieces of innovation. They are getting interconnected and there is convergence between them. It is this combinatorial technological changes and innovation which will radically improve human experience that cannot be predicted by traditional extrapolation method.

As far back as May 2014, Stephen Hawking along with others had written that creation of a true thinking machine would be the biggest event in human history. It can outsmart financial markets; out-invent human researchers, out-manipulate human leaders and develop weapons that we cannot even imagine. To dismiss all these as science fiction would be a mistake. James Rodman Barrat, a prolific writer had conducted a survey on Artificial General Intelligence (AGI) and asked 200 top notch computer scientists and technologists to indicate when AGI would be achieved. The response indicated that 42% believed by 2030, 25% by 2050 and 20% by 2100. The huge momentum that this field has generated can be seen by deep pocketed corporations like Google, Facebook, Amazon etc. adopting artificial intelligence (AI) as central to their business. Not only the corporate sector but even nations are in the race to lead in the field of AI because of the huge potential and possibilities. The future is going to be very different. Not that we know exactly how this change is going to happen or when, but surely the twenty-first century will witness one of the most transforming and thrilling periods in history. It will be an era in which the very nature of what it means to be human will be both enriched and challenged, as our species would break the shackle of its genetic legacy and achieve inconceivable heights of intelligence, material progress, and longevity. It will be an era of Singularity, an era in which our intelligence will become increasingly non-biological and many times more powerful than it is today.

This book is about technologies which are ushering in singularity. In astrophysics, a singularity refers to the point within a black hole where normal laws of physics break down. Gravitational force is so intense there, light itself is unable to escape its grasp. Technological singularity will be similar. Ray Kurzweil, Google's Director of Engineering and a noted futurist, has traced a well-thoughtout path to technical singularity in his book *The Singularity is Near*. He said singularity is that point in time when all the advances in technology, particularly in artificial intelligence (AI), will lead to machines that are smarter than human beings and predicted that singularity will happen sometime around 2045. Bigwigs like Stephen Hawking, Elon Musk, and even Bill Gates have warned about the darker side of AI. But singularity is also an opportunity for humankind to improve. One thing is sure that the future will see an unparalleled human-machine synthesis with a potential to exemplify all things that we value in humans to a great degree. Discerning students and professionals will like to think about these exciting times filled with mind-boggling technologies and become passionate enough to pursue these as their career.

The idea of writing this book did not come up suddenly while sitting under an apple tree and an apple dropping on the head. It is the result of interaction with large number of students from schools, colleges and universities and alumni, about the shape of things to come in engineering and technology in the foreseeable future. Apparently, the interest in engineering has taken a beating, in recent times, as can be seen by the dwindling admissions in engineering colleges. IITs and NITs are doing no better. Drop in manufacturing activities and low job creation can be attributed to the temporary slowing down of the engineering sector. Government of India is taking lot of initiatives like Make in India programme to counter this downward trend and the country is on its way to become a future global power house. The Indian economy expanded 7.2 percent year-on-year in the last three months of 2017, well above an upwardly revised 6.5 percent advance in the previous period, beating market expectations of 6.9 percent growth. The World Bank on 14 March 2018 said, India's GDP growth rate will return to 7.5% in two years' time. On this positive note, this book will add a wow factor by discussing technologies which will lead the way to technological singularity.

As already stated, exponential growth and convergence of Genetics, Robotics, Artificial Intelligence and Nanotechnology are going to pave the way for machine intelligence to compete with human intelligence for a place in posterity.

- The genetics revolution will allow reprogramming our own biology.
- The nanotechnology revolution will allow manipulation of matter at the molecular and atomic scale. It will accelerate reprogramming of human biology.
- The robotic revolution in conjunction with artificial intelligence and its variants like machine learning and deep learning will allow creation of powerful non-biological intelligence.

Each one of the above factors is powerful in its own right and we are already experiencing them in some capacity. But it is the convergence of these technologies which will bring in revolution. Elon Musk of Tesla and SpaceX fame, has launched a company called Neuralink Corp which would help prevent humans from becoming "house cats" to AI. Tiny electrodes will be implanted into the brain that may one day upload and download thought. Its primary purpose is to optimize mental output through a brain-computer interface, allowing the human brain to effortlessly access the internet and, thus, keep up with it and someday merge with artificially intelligent systems. There are others like Bryan Johnson of Kernel, who is planning to apply Silicon Valley playbook to neuroscience. As the trend continues, carbon and silicon-based intelligence will merge to form a single global consciousness. An ultra-intelligent machine could design even better machines. Thus, super-intelligence may become an unstoppable power because of its intellectual superiority and the technologies which can out-think and out-smart the human. It is crucial that a robot be provided with human-friendly motivations or "friendly AI". Engineers and designers will have huge moral and ethical responsibilities to ensure humanity is not harmed by the newly acquired power by machines. The book also identifies educational institutions and research organizations which are working in these cutting edge areas.

Fabiola Gianotti, a particle physicist and the Director General of CERN, is the woman in charge of the Large Hadron Collider as well as other Big Science projects. While addressing the world economic forum in January 2018 she said, "We need to break the cultural silos. Too often people put science and the humanities, or science and the arts, in different silos. They are the highest expression of the curiosity and creativity of humanity." This brings us to liberal arts education which is by nature broad, diverse and inclusive. A student gets opportunity to study a bouquet of subjects like history, mathematics, computer science, classical literature etc. as part of the same curriculum and develop skills in creative thinking, critical thinking, sensitivity and capacity for life-long learning. These will be assets for professionals working in the field genetics, nanotechnology, super intelligence and making of friendly AI.

May this book ignite the imaginations of discerning students and professionals to find their winds of passion in the technologies which will usher singularity.

CHAPTER 1
SINGULARITY

On Feb. 15, 1965, a high school student named Raymond Kurzweil appeared as a guest on a game show called *I've Got a Secret*. After being introduced by the host, Steve Allen, he played a short musical composition on a piano. The idea was that Kurzweil was hiding an unusual fact and the panellists, a comedian and a former Miss America, had to guess what it was. The comedian guessed it right, the music was composed by a computer. Kurzweil then demonstrated the computer, which he had built himself, a desk-size affair with loudly clacking relays, hooked up to a typewriter. He would spend much of the rest of his career working out what this demonstration meant, creating a work of art, usurped by a computer – a line blurred between organic intelligence and artificial intelligence. That was Kurzweil's real secret, and back in 1965 nobody guessed it, not even him. Forty six years later Kurzweil now believes that a moment in time is approaching when computers will become intelligent and not just intelligent but will surpass human intelligence. He calls that as *Singularity,* a point in time when artificial intelligence will surpass biological intelligence. Then humanity, our bodies, our minds, our civilization, will be completely and irreversibly transformed. Singularity will no more be a science fiction but a reality. The pace of change is accelerating, and new potentially disruptive technologies are on the horizon. There is a need to develop a strong understanding and perspective on such technologies. We need to survey new innovations, study and forecast their pace, gauge the implications, and adopt new tools and strategy to chart a passionate course in life. Nurturing curiosity is the first step to understanding technological change. We have got to know what's in the pipeline. It isn't all about reading, but also taking a plunge where the change is happening. Whatever our strategy, the goal should be to develop a healthy obsession with technology. Exploration of fresh perspectives outside traditional work environments and giving ourselves permission to see how these new ideas tickle our passion is important. It's not a fringe idea; it's a serious hypothesis about the future of life on Earth. It's an idea that demands careful evaluation. (Lev Grossman in Times dated 10 Feb 2011) We will, hence, devote some time with such astounding changes.

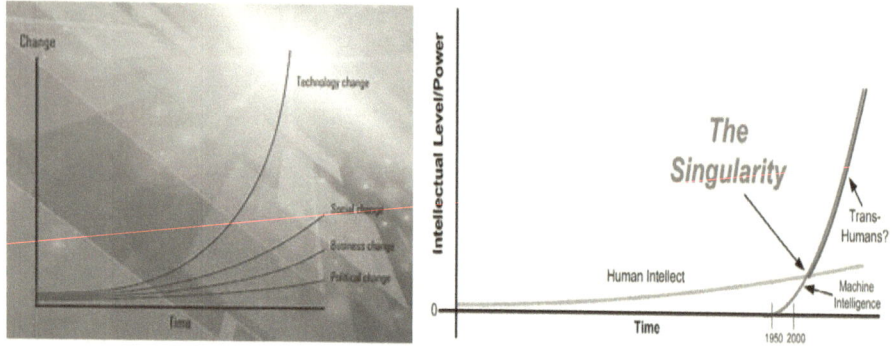

Google Images for Accelerating Growth of Technology

One of the most challenging aspects is that few technology arenas are quickly moving from a linear to an exponential pace, from a growth of 10% per year to more than double annually. That is another world altogether. For example, machine learning and big data are finally reaching critical mass after more than twenty years of being right around the corner. We have begun to understand implications of applications like speech and image recognition. Along with the disruptive change in one technology, far more exciting possibilities are unfolding as multiple arenas are converging. Hod Lipson, Professor of Mechanical Engineering and Data Science at Columbia University and co-author of *Driverless: Intelligent Cars and the Road Ahead*, elaborates on this convergence when he says, "AI is the engine, but big data is the fuel. They need each other." This convergence paired with an accelerating pace makes for surprising applications. Ray Kurzweil, now Director of Engineering at Google and leading futurologist, states "in the 2030s, we are going to send nano-robots, via capillaries, into the brain that will provide full immersion virtual reality from within the nervous system and will connect our neocortex to the cloud. Just as we can wirelessly expand the power of our smart phones 10,000-fold in the cloud today, we'll be able to expand our neocortex in the cloud." Let's understand this statement. He is talking about directly plugging the brain into the internet and upgrading the intelligence and memory capacity by orders of few magnitude. Such technologies may sound fanciful or even deceptive but they are actually becoming disruptive. To cite another example, CRISPR/Cas9, a revolutionary gene-editing technique, is the new favourite as well as villain of genetics research which almost certainly will play a major role in the future. So if these fields of disruptive technological progress double every 18 months, by 2030 we will have a 1,000-fold improvement over today. And what does a future one thousand times better will look like? Before we go on to see few such technologies in the subsequent chapters, let us first get an overview of singularity in general.

1.1 WHAT IS SINGULARITY

The term "Singularity" has come originally from the field of theoretical physics. It means a point where some property is infinite. For example, at the centre of a black hole, according to classical theory, the density is infinite because a finite mass is compressed to a zero volume. Hence it is a singularity.

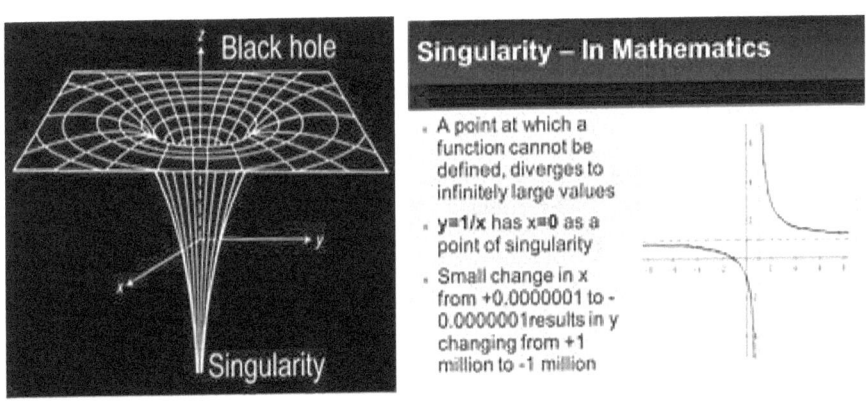

Google Images of Singularity

It also refers to a mathematical term where the value or range of values of a function for which a derivative does not exist. By analogy, a singularity in human history would occur if exponential technological progress due to dramatic changes would bring an end to human affairs. The most basic human values like sanctity of life, pursuit of happiness, freedom of choice would be superseded. What we are most interested in, however, is the definition of Singularity as a technological phenomenon i.e. the Technological Singularity. It has definitely become a mainstream topic. Different technologist and futurists have defined Technical Singularity in different flavors. Let us see a few definitions.

1.2 TECHNOLOGICAL SINGULARITY

Vernor Vinge introduced the term technological singularity in January 1983 in the Omni magazine and talked of the creation of intelligent machines. "We will soon create intelligences greater than our own. When this happens, human history will have reached a kind of singularity." (www.33rdsquare.com) He later developed this concept further in his essay *Coming Technological Singularity (1993)* wherein he said "Within thirty years, we will have the technological means to create superhuman intelligence. I think it's fair to call this event a singularity." Vinge's singularity would occur when computers are developed to be conscious and superhumanly intelligent.

Ray Kurzweil is one of the most talked about proponent of singularity. He has generally embraced Vernor Vinge's term but put it differently. In his book *The Singularity Is Near* Kurzweil has defined the technological singularity as "a future period during which the pace of technological change will be so rapid, its impact so deep, that human life will be irreversibly transformed. Although neither utopian nor dystopian, this epoch will transform the concepts that we rely on to give meaning to our lives."

James Martin, a world-renowned futurist, computer scientist, author, lecturer defines singularity as "a break in human evolution that will be caused by the staggering speed of technological evolution."

Sean Arnott has said "The technological singularity is when our creations surpass us in our understanding of them vs their understanding of us, rendering us obsolete in the process."

1.3 CONSEQUENCES OF SINGULARITY

It may not be possible to correctly predict the behavior of these smarter-than-human intelligences with which we might one day share the planet. What are the consequences? Should we fear the singularity or should we welcome it? The possibilities of technological singularity are both an existential risk and an existential opportunity. It is a risk because it opens up possibilities like enabling a highly contagious drug resistant virus to be genetically engineered resulting in a catastrophe. Advanced AI can also be an existential risk. On the other side of the spectrum, technological singularity can also be an existential opportunity. The capability to engineer the mind opens up the possibility of transcending our biological heritage. Our body is fragile, vulnerable to disease, damage and decay. But if we can acquire the means to rebuild it from scratch, then we will be looking at an unlimited extension of consciousness, being termed as "transhumanism". If we can rebuild it, why not totally re-engineer the brain, leading to a more radical form of cognitive enhancement and re-organisation!

1.4 WILL SINGULARITY BE REAL

Futurists like Vernor Vinge and Ray Kurzweil have argued that the world is rapidly approaching a tipping point, namely Singularity, where the accelerating pace of smarter and smarter machines, having super human intelligence, will outperform all human capabilities. When greater than human intelligence drives progress, that progress will be much more rapid. From the human point of view, this change will be a throwing away of all previous rules. Our life experiences would be altered in an unimaginable way. Kurzweil envisions a future in which developments in medical nanotechnology will allow us to download a copy of our individual brains into these superhuman machines, leave our bodies behind, and, in a sense, live forever. Immortal beings!!

We do take that singularity might occur one day, though it may take some time. Kurzweil has however predicted, by working through a set of models and historical data, that singularity will arrive around 2045 primarily based on Laws of Accelerating Return. But if the singularity is to arrive by 2045, it will take fundamentally unpredictable breakthroughs, and not merely because of the Law of Accelerating Returns. The rapid advances in technology have to occur not only in computer hardware but also in software. We would need to build smarter and more capable software programs based on understanding of the foundations of human cognition. This means not just knowing the physical structure of the brain, but also how the brain reacts and changes, and how billions of parallel neuron interactions can result in human consciousness and original thought. The complexity of the brain is awesome. Every structure has been precisely shaped by millions of years of evolution and environmental factors. More we learn, the more we realize there is more to know. But of course comprehensive understanding of the brain is possible at a future date consequent to the arrival of powerful new theories and it will then decide the time frame to achieve the singularity.

When Ray Kurzweil published *The Singularity Is Near* in 2006, many scoffed at his outlandish predictions. The fact that singularity a serious hypothesis of future life and not a science fiction or a fringe idea can be seen by the fact that it has caught attention in a big way from the large number of singularity conferences happening around the globe. Founding of the *Singularity University* in 2008 by Peter Diamandis and Ray Kurzweil is another mile stone in that direction. Today, Kurzweil's ideas don't seem outlandish. Google's Deep Mind (DeepMind is a world leader in AI. Company was founded in 2010 in London and was subsequently acquired by Google in 2014. www.deepmind.com) created AlphaGo. This AI recently beat legendary Go world champion Lee Sedol who is a professional Go of 9 dan rank and is one of the strongest player in the history of the world. IBM's Watson is expanding horizons in medicine, financial planning and even cooking. Self-driving cars are expected to be on the road by 2020. Technology seems to be accelerating faster than ever before. Greg Satell wrote in Forbes, June 2016 issue, about three reasons which give credence to happening of singularity in near future.

(a) **Technology is beating Moore's Law:** In the mid-1970, Gordon Moore, a leading inventor of integrated circuits and later Chairman of Intel, observed that we could squeeze twice as many transistors into integrated circuits every twenty four months. Given that the electrons would consequently have less distance to travel, circuits would run faster, providing a boost to the computing power. For the last 50 years, the technology industry has been driven by Moore's Law. Now we are approaching the theoretical limit as transistors can be shrunk to a limit after which quantum effects between atoms cause them to malfunction. While the process of cramming more transistors on silicon wafers is slowing down, variety of new ways, such as quantum computing, nanotubes, DNA computing, spintronics, neuromorphic chips and 3D stacking are speeding up performance. Computational capacity will thus continue to grow exponentially long after Moore's Law ends and will eventually rival the raw computing power of the human brain.

(b) **Robots are Doing Human Jobs:** GM assembly line in 1962 used the first industrial robot, Unimate, for welding of auto parts. Automation, since then, has progressively been used in various spheres of life. Beside industrial robot, machines are increasingly being used to aid human being like automatic teller machines in the 1970's to the autonomous Roomba vacuum cleaner in 2002. Rethink Robotics makes robots like Baxter and Sawyer, which can work safely around humans and can learn new tasks in minutes. Military robots are becoming commonplace on the battlefield and soldiers are developing emotional bonds with them, even going as far as to hold funerals for their fallen android brethren. And lest one thinks that automation only applies to low-skill, mechanical jobs, robots are also invading the creative realm. One book written by a machine was even recently accepted as a submission for the prestigious Hoshi Shinichi Literary Award in Japan. Department of Defense is experimenting with chips embedded in soldier's brains and Elon Musk is thinking of commercializing such technologies. As the power of technology continues to grow exponentially—computers will be more than a thousand times more powerful and robots will take on even more tasks. It is pertinent to mention here that it is not only the exponential growth in hardware which is required to usher singularity but creation of advanced software which will understand human cognition and human brain, not only its physical structure but also how billions of parallel neuron interactions results in human consciousness and original thought.

(c) **We are editing genes:** In 2003, scientists created a full map of the human genome. For the first time, we actually came to know about genes and could begin to track their function. Just two years later, in 2005, the US government started compiling the Cancer Genome Atlas, which allows doctors to target cancers based on their genetic makeup rather than the organ in which they originate. Now, scientists have a new tool at their disposal, called CRISPR, which allows them to actually edit genes, easily and cheaply. It is already opening up avenues to render viruses inactive, regulate cell activity, create disease resistant crops and even engineer yeast to produce ethanol that can fuel our cars. The technology is also creating quite a controversy and raising moral and ethical issues. When one starts editing the code of life, where does one stop? Should we create designer babies, with predetermined eye color, intelligence and physical traits?,

1.5 TECHNOLOGIES LEADING TO SINGULARITY

Ray Kurzweil pointed out that as technology accelerates at an exponential rate, progress would eventually become virtually instantaneous, a singularity. Further, as computers advanced, they would merge with other technologies, namely genomics, nanotechnology and robotics. Most of us have grown up with the internet, smartphones, tablets and apps for everything. These new technologies have altered our lifestyle. But have we reached a dead end? What else could turn up that could make our lives much more different? Faster computers? More gadgets? It has to be much more than that. Technologies have embarked on an exponential growth curve. Ten years hence we will look back and wonder how we could have lived with such primitive technologies. The gap will be huge.

Of all the technologies riding the wave of exponential progress, the first half of the twenty-first century will be characterized by revolution in three overlapping technologies—**genetics, nanotechnology, and robotics & artificial intelligence** (AI), and their many variations. A very brief introduction is given below and more details will follow in subsequent chapters;

(a) **Genetics:** Genetics has progressed tremendously over the last 15 years. From the sequencing of the first full human genome in 2003, we have now entered the era of personal genomics which will allow us to programme our own biology. Understanding the information processes underlying life, and reprograming our biology will help in achieving the virtual elimination of disease, dramatic expansion of human potential, and radical life extension. Creation of "superhumans" also known as transhumanism and approaching the age of genetically enhanced humans will be the revolution in Genetics.

(b) **Nanotechnology:** This revolution will enable us to redesign and rebuild at molecular level and atomic scale. In accordance with law of accelerating returns, key features of many electronics and even mechanical technologies will be in the nanotechnology range, generally considered to be less than 100 nanometers. Scientist at McGill University in Montreal have demonstrated a nanopill of 25–30 nanometer range which will pass through the cell wall and deliver medicine directly to targeted structure inside the cell. One of the most important future nanorobotics application in future, could be merger of biological and non-biological intelligence.

(c) **Robotics & AI:** But the most powerful impending revolution is the robotic and AI which will allow creation of a greater than human non-biological intelligence. Starting with cognitive computing, IBM's Watson is already capable of reading a million books a second and answering question posed in natural language. Most importantly, Watson can provide better medical diagnostics than any human medical doctor, give financial advice, as well as generate or evaluate all kinds of scientific hypotheses. Computer power increases in average 100 folds every 10 years, which means 10,000 folds after 20 years, and 1 million folds after 30 years. Imagine what computers will be able to do in foreseeable future. Peter Diamandis, co-founder of Singularity University, chairs The X Prize Foundation, managing incentivized competitions to bring about radical breakthroughs for the benefit of humanity. In one of the current competitions, the Nokia Sensing XCHALLENGE, a smartphone-like device, tested vitals like cholesterol, blood pressure, heart rate, analyse DNA for genetic risks, diagnose medical conditions, and predict potential diseases or the likelihood of a stroke. Apple is supposedly working on an iPhone with similar capabilities. Google is working on AI that will be able reading and understanding any document, learn the content of all books and be able to answer any question. This omniscient AI will eventually become people's first source of knowledge, replacing schools, books and even human interactions.

(d) **3D Printing:** 3D printing is the biggest upheaval in manufacturing since the industrial revolution. One can print objects in three dimensions as well as practically in any material like metals, concrete, fabrics, and even food. High-quality 3D printers can copy electronic chips in the tiniest detail and have a functional chip. High-tech vehicles like the Koenigsegg's One:1 (the world's fastest car) or EDAG's Genesis are already being made by 3D Printing. Even houses will be 3D-printed, for a fraction of the costs of traditional construction. In the near future we won't need to go shopping to buy new products. We will just select them online, perhaps tweak a bit their design, size or colour to our tastes and then just 3D print them at home. The good news is that it will considerably reduce our carbon footprint by cutting unnecessary transport from distant factories. 3D printing is also good news for medicine. Doctors can now make customized prosthetics, joint replacements, dental work and hearing aids. In the future it will be possible to regrow limbs or organs directly on a person, as if the body was simply healing itself. Combing 3D printing and stem cell regeneration paves the way to the printing of human organs, a field known as **Bioprinting**

(e) **Bioengineering:** Convergence of advances in robotics, AI, 3D printing and nanotechnologies is leading to the field of bioengineering. Human cyborgs are not science-fiction anymore. Soon microchips will allow bionic humans to see and hear better than humans in their natural state. Equipped with one of these, humans will be able to see ultraviolet and infrared, hear ultrasound like animals, echolocate like bats, and perhaps even eventually understand animal languages, including the marine vocalization. The potential for improvements is unlimited. We are on the verge of developing telepathic abilities by connecting the brains of two individuals and connecting them through the internet. Possibilities are enormous.

1.6 WHY ARE WE SKEPTICAL ABOUT EXPONENTIAL FUTURE?

> We think in straight lines when we think of history. When we imagine the progress of the next 30 years, we look back to the progress of the previous 30 as an indicator of how much will likely happen. When we think about the extent to which the world will change in the 21st century, we just take the 20th century progress and add it to the year 2000. It's intuitive for us think linearly, when we should be thinking exponentially. Kurzweil suggests that the progress of the entire 20th century would have

been achieved in only 20 years at the rate of advancement in the year 2000—in other words, by 2000, the rate of progress was five times faster than the average rate of progress during the 20th century. He believes another 20th century's worth of progress happened between 2000 and 2014 and that another 20th century's worth of progress will happen by 2021, in only seven years. A couple decades later, he believes a 20th century's worth of progress will happen multiple times in the same year, and even later, in less than one month. All in all, because of the Law of Accelerating Returns, Kurzweil believes that the 21st century will achieve 1,000 times the progress of the 20th century.

➤ Even a steep exponential curve seems linear when we only look at a tiny slice of it. And, exponential growth isn't totally smooth and uniform. Kurzweil explains that progress happens in "S-curves" which goes through three phases. It is slow growth in the early phase of exponential growth, then the explosive phase and finally levelling off. So someone looking at the levelling phase will miss the point.

➤ We base our ideas about the world on our personal experience, and that experience has ingrained the rate of growth of the recent past in our heads as "the way things happen." We're also limited by our imagination, which takes our experience and uses it to conjure future predictions—but often, what we know simply doesn't give us the tools to think accurately about the future.

Being truly logical and expecting historical patterns to continue, should tell us that much more will change in the coming decades than we intuitively expect. Logic also suggests that if the most advanced species on a planet keeps making larger and larger leaps forward at an ever-faster rate, at some point, they'll make a leap so great that it completely alters life as they know it and the perception they have of what it means to be a human—kind of like how evolution kept making great leaps toward intelligence until finally it made such a large leap to the human being that it completely altered what it meant for any creature to live on planet Earth. A general awareness about what's going on today in science and technology will make us see a lot of signs quietly hinting that life as we currently know it cannot withstand the leap that's coming next.

1.7 ETHICS IN SINGULARITY

One of the first to consider seriously the idea of the singularity was the philosopher and computer scientist I.J. Good who said in 1965, "Let an ultraintelligent machine be defined as a machine that can far surpass all the intellectual activities of any man however clever. Since the design of machines is one of these intellectual activities, an ultraintelligent machine could design even better machines; there would then unquestionably be an "intelligence explosion", and the intelligence of man would be left far behind. Thus, the first ultraintelligent machine is the last invention that man need ever make." This is the singularity: an explosive ride of increasing intelligence. Since the super-intelligence may become unstoppably powerful because of its intellectual superiority and the technologies, it is crucial that it be provided with human-friendly motivations or "friendly AI". Singularity Institute (www.singinst.org) writes that a "Friendly AI" is an AI that takes actions that is, on the whole, beneficial to humans and humanity; benevolent rather than malevolent; nice rather than hostile.

The idea of approaching a technological singularity is both exciting and scary. While the prospects of technologies that are hundreds of times more powerful than what we have today will open up completely new possibilities, there are also inherent dangers. How autonomous should we allow robots to become? Which genes are safe to edit and which are not? Technologies like AI, Genetics, Nano will inevitably give rise to new ethical, social, and moral questions that we have never faced before. Rather than bury our heads in the sand, we must explore the full range of potential consequences of whatever is underway or still to come. We can add AI to kids' toys, like Mattel's Hello Barbie or use cutting-edge gene editing technology like CRISPR-Cas9 to select for preferred gene sequences beyond basic health. But just because we can do something doesn't mean we should. Elon Musk, Stephen Hawking, Steve Wozniak, Bill Gates, and other well-known names in science and technology have expressed concern through media and open letters about the risks posed by AI. Microsoft's CEO, Satya Nadella, has even argued tech companies shouldn't build artificial intelligence systems that will replace people rather than making them more productive. Exploring unintended consequences goes beyond having an alternate plan for when something goes wrong. So the creator of such technologies must be highly responsible world-citizens with holistic view of morality and responsibility per se. They should understand how their decisions will impact communities and the environment. For them:

The question isn't: "If we build it, will they come?"

The question is: "If we can build it, should we?"

Engineers and designers have first a moral and ethical responsibility and then creativity and superior domain knowledge. The chapter on Liberal Education will discuss how it can provide a holistic basis of education to produce responsible world-citizen.

REFERENCES

1. **Singularity University (https://su.org):** S U was jointly founded by Dr. Peter H. Diamandis and Dr. Ray Kurzweil in 2008 with a mission to educate, inspire and empower leaders to apply exponential technologies to address humanity's grand challenges. The university acts as a platform for visionaries, technologists, leaders and resources to come together and becomes a launchpad for new thinking, innovation, powerful entrepreneurs and doers to solve exponential challenges. Some of their programmes are:
   (a) Executive Program: This is a week-long course in Silicon Valley that examines how key converging technologies will shape our future and explores ethical leadership in a world of rapid change.
   (b) Global Solutions Program: GSP is a residential experience at Singularity University's campus for dreamers and inspired solvers with unique skill sets, experiences, and passions, and empower them to develop moon shot innovations by leveraging exponential technology and entrepreneurial tools.
   (c) Exponential Foundations Series Online: Over nine weeks, participants are guided to think exponentially and develop the skills for building an abundant future.
2. Accelerating Future: https://acceleratingfuture.com/
3. FERN (Foresight Education and Research Network): https://www.fernweb.org/
4. Future of Humanity Institute: https://www.fhi.ox.ac.uk/
5. Next Big Future: https://www.nextbigfuture.com/
6. Singularity Hub: https://singularityhub.com/

7. Singularity is Near: www.singularity.com
8. Forbes: https://www.forbes.com/sites/gregsatell/2016/06/03/3-reasons-to-believe-the-singularity-is-near/#42584bcb7b39
9. Singularity is Near wiki: https://en.wikipedia.org/wiki/The_Singularity_Is_Near
10. AI Revolution by Tim Urban: https://waitbutwhy.com/2015/01/artificial-intelligence-revolution-1.html

CHAPTER 2
LIBERAL ARTS EDUCATION

2.1 WHY LIBERAL ARTS

One may wonder why a book on exponential technologies should include Liberal Arts as a topic for discussion. Well that is a reasonable question—Why *should* a budding student of high end technology study history, literature, philosophy, music, art, or any other subject outside of curriculum in STEM i.e. Science, Technology, Engineering and Mathematics? Why should anyone study a subject that does not, apparently, help them getting a job in technologies of the future?

2.2 QUOTES FROM WORLD ECONOMIC FORUM

Few quotes on education and learning as discussed in World Economic Forum, Davos in January 2018 will answer the question posed above:
- The future of work is going to look very different, as automation and Artificial Intelligence make many manual, repetitive jobs obsolete.
- According to the McKinsey Global Institute, robots could replace 800 million jobs by 2030, while the World Economic Forum suggests a "skills revolution" could open up a raft of new opportunities.
- "If we do not change the way we teach, 30 years from now, we're going to be in trouble," said Jack Ma, founder of Alibaba Group, China's e-commerce giant.
- The knowledge-based approach of "200 years ago", would "fail our kids", who would never be able to compete with machines. Children should be taught "soft skills" like independent thinking, values and team-work, Jack Ma said.

- "Anything that is routine or repetitive will be automated," said Minouche Shafik, Director of the London School of Economics, in a session on Saving Economic Globalization from Itself. She also spoke of the importance of "the soft skills, creative skills, Research skills, the ability to find information, synthesise it, make something of it."
- Fabiola Gianotti, a particle physicist and the Director General of CERN, the woman in charge of the Large Hadron Collider as well as other Big Science projects, says music is as important as maths. "We need to break the cultural silos. Too often people put science and the humanities, or science and the arts, in different silos. They are the highest expression of the curiosity and creativity of humanity," she said in a session on education. "For me, I was a very curious child, I wanted to answer the big questions of how the universe works. My humanities and my music studies have contributed to what I am today as a scientist as much as my physics studies."

 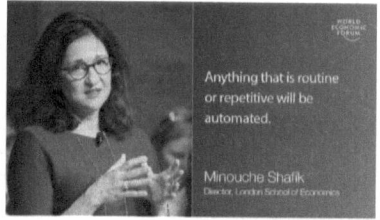

- Helping kids play more "will equip them to be relevant to the workplace and to society," said John Goodwin, CEO of the Lego Foundation and the former chief financial officer for The Lego Group. That may sound self-serving coming from a Lego executive but research shows that play is crucial in establishing the foundations of social, emotional, and academic learning. Dressing up like Batman or building imaginary cities with blocks help young children cultivate creativity, develop emotional intelligence and regulation, and build empathy—the very skills that robots can't replace.
- At Davos, this notion was popular even among those who don't build toys for a living. Kai Fu Lee, a Taiwanese venture capitalist who opened Google's China office and who has worked in artificial intelligence (AI)

for more than three decades, said we need to develop the skills that are unique to humans. "There are four things AI cannot do as well as humans: creativity, dexterity, compassion, and complexity." Empathy, he said, would be paramount. "We have a human responsibility to do this."

Education should not be something that is done at a specific institution for a specific period of time for just getting a degree. It is a life long journey of exploration, self-discovery and liberation driven by intrinsic reward. Effective education is something that is integrated into our everyday experiences. A lifelong learning mind set is essential in order to be employable in future work. A report of World Economic Forum revealed that almost 65% of the elementary school students will be working in an area in the future which do not even exist today.

2.3 SKILLS OF 21ST CENTURY

If we would have asked a farmer just a few years ago what skill their kids would need to thrive, the answer would have been how to milk a cow or farm a field. Now machines are mostly doing these work. With each generation and even within generations, some jobs are disappearing. New industries are being born and some existing ones dying out because of disruption. So the question can be resolved by identifying what skills are to be taught to the current generation kids so that they can keep pace with accelerating technology.

Finland recently shifted its national curriculum to a new model called "phenomenon based" approach. They will replace the traditional class room subjects with an approach which highlights four Cs, communication, creativity, critical thinking and collaboration, which are central to working in teams and a reflection of hyper connectivity of the world. David Hill, Singularity Hub Editor-in Chief has endorsed the four Cs being adopted by Finland. Successful entrepreneurs across the globe are talking of three additional soft skills, adaptability, resilience and grit, which can be integrated with four Cs. Esteban Bullrich, Buenos Aires' Minister of Education told Singularity University in an interview "I want kids to get out of school and be able to create, to be able to change the world with capabilities they earn and receive through formal schooling". This message is consistent with research highlighted in a World Economic Forum and Boston Consulting Group report titled, *New Vision for Education: Unlocking the Potential of Technology*. The report divides the 21st century skills into three categories as given in image below:

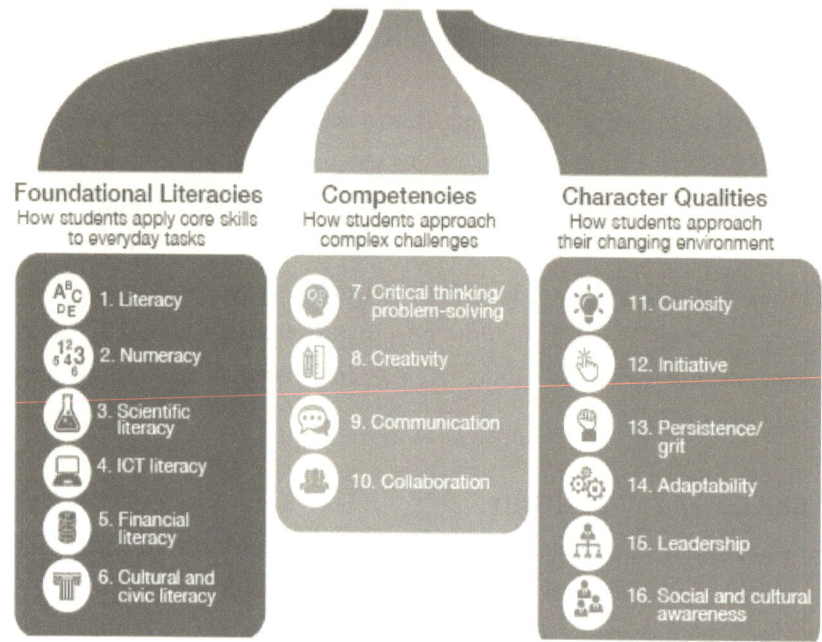

This continuous learning approach, in contrast to degree oriented education, represents an important shift that is desperately needed in education. It is also getting reflected in the future job market where lifelong learning and skill development will keep an individual competitive, agile and valued.

Beside relevance in job market, coming of singularity will also impose a huge moral, ethical and technological responsibility on the maker of technology which will lead the way to that critical point of time. There are pros and cons about happenings of singularity. The optimists envision a future in which artificial intelligence will help rid the world of disease and extend our own lives beyond frail biological limitations. The pessimists, on the other hand, are arguing that, even at the outset, sufficient safeguards must be thought of to protect us from machines who will know more than us, and think faster. Nick Bostrom, the head of the Future of Humanity Institute of UK, says that the superintelligent machines that will walk the earth one day, must be given *our* values. Meaning, they may be artificial, but they would be actors with minds of their own. Hence we can imagine the responsibility of the leaders who will instil or enforce or impose values on artificial intelligence.

2.4 HOW WILL LIBERAL EDUCATION HELP

- A liberal arts education will develop an ordered intellect and teach how to think. The mind is like a muscle; exercise makes it stronger and more able towards grasping ideas and intellectual work. Exercising the mind in one area, whether literature or sociology or accounting, will strengthen it for learning in other areas as well. Thinking has its own grammar, its own orderly structure and set of rules for good use. Many subjects help the student to develop an ordered mind, and each subject contributes in a slightly different way. A careful study of music or logic or good poetry will irresistibly demonstrate the structure of thought and knowledge and intellectual movement and will create the habit of organized thinking and of rational analysis. Once habit of good thinking is developed, one will be able to perform better in any job, and also lead a happier because of the knowledge of organized solutions, of hierarchical procedures, of rational sequences that can be applied to any endeavor.
- The diverse body of knowledge one gains from a liberal arts education, together with the tools of examination and analysis will enable development of one's own opinions, attitudes, values, and beliefs, based upon one's worthy apprehension, examination, and evaluation of argument and evidence. Diverse studies will permit to see the relations between ideas and philosophies and subject areas and to put each in its appropriate position. Good judgment depends upon a thoughtful and rather extensive acquaintance with many areas of study and the ability to think independently, in the face of pressures, distortions, and overemphasized truths.
- The world becomes understandable thorough knowledge of a wide range of events, philosophies, procedures, and possibilities. No longer will unexpected or strange things be merely dazzling or confusing. How sad it is to see a mind educated in only one discipline completely overwhelmed by a simple phenomenon. Abraham Maslow, the philosopher had said, "When the only tool you have is hammer, you tend to look at every

problem as a nail." A wide ranging education, covering everything from biology to history to human nature, will provide many tools for understanding. Thus, to see a complex situation fits into the whole, liberal education is not merely desirable but necessary.

- General knowledge enhances creativity. Knowledge of many subject areas provides a cross fertilization of ideas, a fullness of mind that produces new ideas and better understanding. The interactions of diversified knowledge are so subtle and sophisticated that their results cannot be predicted. When Benjamin Franklin flew a kite into a storm to investigate the properties of electricity, he did not foresee the wonderful inventions that future students of his discoveries would produce—the washing machines, microwave ovens, computers, radar installations, electric blankets, or television sets. Nor did many of the inventors of these devices foresee them while they studied Franklin's work.
- Life itself is wholesome, not divided into majors. We suffer every day from consequence of not recognizing the fact that most endeavors require more knowledge than one specialized field. A coder of machine language with knowledge only in Python and R will never fully understand the consequence of artificial intelligence surpassing biological intelligence. But a study of biology, sociology, logic and fine arts will give him the intellect to grasp the enormity of the singularity.

2.5 LIBERAL ARTS IN INDIA

For a long time in India, degree in science, engineering, medicine, commerce, art simplified relatively rigid disciplines. The idea of a course where a student could study biology and classical literature as part of the same curriculum seemed absurd. Traditionally, the U.S. was the first port of call for liberal arts aspirants. However, the number of universities in India, offering Liberal Arts has grown, as also the number of applicants. Indian institutions are offering quality curricula at much lower cost which is drawing attention of parents and aspiring students and many students are opting to stay back. The career prospects in India after graduation are becoming better.

Contrary to popular belief, liberal art is not only about the fine arts or the performing arts. In fact, the use of the word 'art' is often narrow. For the ancient

Greeks, mathematics, grammar and music were all arts. Art is something made, something fashioned by man. By this understanding, all sciences are 'art' as well, since these scientific knowledge systems are an achievement of man. So a Biology course in a liberal arts program will have a core Biology syllabus comparable to a B.Sc program. The only difference is the approach taken in teaching the course. The approach is one of relevance, interconnectedness and vitality.

As already brought out, merely getting placed in a job is not the sole aim of a well-rounded education. Job opportunities are important, but they aren't the be all and end all. The purpose of an education is to equip with the skills needed to live life to the optimum, within the environment. To be able to do this, one needs to have certain vitality in addition to being multi-dimensional. A liberal arts education, in its open flow of subjects and courses through the arts, sciences and business, offers precisely these two qualities. What distinguishes a liberal arts program from other 'traditional' courses is that every student would have to study subjects from across the academic spectrum. One could major in Biology and pick Philosophy as a minor. Alternatively, one could major in Mathematics or Business Studies while simultaneously working on a History minor. Exposure to these diverse fields is what has made liberal arts the chosen field of study. Diversity is not looked at as an 'add-on' in the liberal arts environment. Rather, it is a necessity. In a vibrant and dynamic world, the need is for vibrant, dynamic but compassionate and sensitive individuals. The environment is becoming complex day by day and hugely interconnected. This will necessitate comprehension of multidimensional factors for holistic decision making. Only professional expertise will not be good enough so long as it is not coupled with ethical and moral concern about the wellbeing of humanity. This performance is no doubt enhanced by an education that broadens the scope of the professional's thought processes and analytical capabilities. A liberal arts education is becoming a necessity.

Quite a few institutions in our country are offering Liberal Education programme in some format or other. Brief details of some of these institutions are given in succeeding paragraphs.

Ashoka University (https://www.ashoka.edu.in/): Ashoka University, in Sonepat, Haryana, is one of the leading Liberal Arts universities in India. It is fully residential. The university prepares students to be ethical leaders through academic research, and experiences in solving real-life issues by critical thinking and effective communication.

> **Admission Process:** Students are short listed based on SAT/ACT score, academic performance in school, essay writing followed by aptitude test and personal interview.

> **Courses Offered:** For a Bachelor's degree, students must pass courses that are broadly divided into three categories viz Foundation Course, Major Course and Co-Curricular Course. Majors are in nine subjects and Interdisciplinary Majors in a further eight subjects, thereby offering students a total of 17 options to Major. Minors are from among 15 courses. Foundation courses comprise of Critical Thinking 1, Critical Thinking 2, Introduction to Mathematical Thinking, Principles of Science, Mind and Behaviour, Literature and the World, Social and Political Formation, Indian Civilizations and Great Books. In every semester, Ashoka University offers several 2-credit Co-Curricular Courses in Performing Arts, Visual Arts, and Language. Co-Curricular Courses are a mixture of practical training in an art form (e.g. dance, theatre, music, painting, ceramics) or language (e.g. French, Latin, Urdu) and critical analysis of the aesthetic, social, political, economic, and historical dimensions of the same. Students have to take two Co-Curricular Courses as part of their graduation requirement. Broad spectrum of this foundation course is sufficiently indicative of the holistic development of the students

Flame University (www.flame.edu.in): Flame University of Pune is built on a vision to build an aspirational destination for students and faculty, to push the design and nature of studies and to create a societal upgradation particularly in the fields of liberal education and leadership. For building overall personality, there is an effective combination of conventional classroom teaching with seminars, tutorials, workshop, projects and field trips in the curriculum. Dr. Devi Singh, Vice Chancellor of FLAME, says "We are aware of the need to align our courses and curriculum to the needs and demand of market".

> **Admission Process:** Flame University conducts an entrance exam, FLAME Entrance Aptitude Test (FEAT) to admit students to various

courses. During the admission process, your overall profile is considered. This is followed by personal interview

- **Courses Offered:** FLAME University offers undergraduate, post graduate and doctoral courses. Three-year under graduate programme in B.A., B.Sc. and BBA and endeavors to deliver an education rooted in life-skills through core courses in Critical Reasoning, Academic Writing, and IT Literacy. Rather than limiting students to the three streams of Arts, Science and Commerce, FLAME allows students to take courses across areas such as Physical and Natural Sciences, Humanities, Social Sciences and Fine and Performing Arts. These expose students to multiple disciplines, and also help in making an enlightened choice about specialization. FLAME has also established the practice of including uniquely-crafted experiential learning programs alongside the academic curriculum, which is innovative and ground-breaking in ourcontext. Through this, all students have access to a set of structured opportunities for personal growth and professional developmentthrough The Development Activities Program (DAP), the Discover India Program (DIP) and the Summer Internship Program (SIP).

Symbiosis School of Liberal Arts (https://www.ssla.edu.in/): SSLA is part of Symbiosis International University is located in Pune. This University is among the oldest Liberal Arts universities in India, and it has adapted the international ideology of Liberal Arts education. Symbiosis aims to promote ethical and value-based learning and cross-cultural sensitization. Symbiosis School of Liberal Arts offers various four year courses in B.A and B.Sc dedicated to Liberal Arts.

- **Admission Process:** Students have to appear in Symbiosis Entrance Test (SET) for assessment in English, General Intelligence, Quantitative Aptitude, Mathematics, and General Knowledge. Shortlisted candidates are required to appear for PI-WAT (Personal Interaction and Written Ability Test).
- **Courses Offered:** SSLA offers students to study a four year full time BA(Liberal Art) or B.Sc(Liberal Art) over a period of eight semester. 18 core courses, eight general courses, one minor specialization subject of six papers and one major specialization subject of ten papers are covered

in these eight semesters. Degree awarded is dependent on the majoring courses. Regardless of the major and minor field of study, the degree aims to create a strong well rounded foundation. Courses have been designed to provide the students with full range of tools to understand, recognise, critique and appreciate their role in dynamic global environment.

Jyoti Dalal School of Liberal Arts located in Mumbai http://liberalarts.nmims.edu/of NMIMS offers three year course BA(Hons) in Liberal Arts.

Jindal school of Liberal Arts and Humanities located in Sonepat, Haryana, https://www.jgu.edu.in/JSLH offers BA (Hons) in Liberal Arts and Humanities.

Ahmedabad University offers management, engineering and science programmes but now plans to roll out an integrated liberal education programme in 2019. "For us, undergraduate education is about providing a broad and deep education that is premised on interdisciplinary, experiential and research based learning" says Pankaj Chandra, Vice Chancellor, Ahmedabad University.

There is an article in Economic Times dated 21st January 2018 on status of Liberal Arts and Social Sciences in India. Pratap Bhanu Mehta, Vice Chancellor of Asoka University has said "For the younger generation in India, the hook to tap into liberal arts education is not just a degree or profession but also the need to ask important questions about who they are". He also said that jobs of the future would be more about reasoning skills and abilities that make one stand out in the market rather than content. "The focus is now shifting to how good you are with reasoning, mathematical skills and writing skills, and with these you can do any job well." Aditi Parekh, a graduate in social science and humanities from Ambedkar University, Delhi says "One of the best features of the liberal arts structure is that it is designed to help you discover different discipline and gives you the option to shape your courses based on emerging interest".

C. Raj Kumar, Vice Chancellor, O.P. Jindal Global University, Dean of Jindal Global Law School and Director, International Institute for Higher Education Research & Capacity Building, observes, "Historically, Indian universities such as Nalanda and Shantiniketan also encouraged a broad-ranging education aimed at holistic development. However, over time our education system has evolved in a manner which takes away the opportunity for students to discover their interests and passion. The new wave of liberal arts education in India addresses this key need among today's increasingly knowledgeable students who have high expectations from their academic life."

REFERENCES
1. The new wave of liberal art: Hindu, 12 December 2015.
2. This is the purpose of liberal arts: Intercollegiate Studies Institute, 13 October 2017.
3. Why study liberal art: Richard Sigurdson, University College of Cariboo.
4. Liberal Education WIKI: https://en.wikipedia.org/wiki/liberal_education

CHAPTER 3
ROBOTICS AND ARTIFICIAL INTELLIGENCE

ROBOT

A play called R.U.R. (Rossum's Universal Robots) was published by Karel Capek, a Czech author and playwright, in 1920. Capek is credited with coining the word "robot," which in Czech means "forced labour" or "drudgery." There is an interesting story, written by Capek himself, about how he came to use the word robot in his play. While the idea for the play was still playing in his mind, he rushed immediately to his brother Josef, the painter, who was standing before an easel and painting. 'I don't know what to call these artificial workers,' he said. 'I could call them Labori, but that strikes me a bit bookish.' 'Then call them Robots,' the painter muttered, brush in mouth, and went on painting.

3.1 LAWS OF ROBOTICS

The science fiction writer Isaac Asimov wrote many stories about robots. His first collection of 1950, *I, Robot*, consists of nine stories about "positronic" robots. He was tired of science fiction stories in which robots such as Frankenstein's creation were destructive. Asimov's robots had "Three Laws of Robotics" hard-wired into their positronic brains and these are:

- **1ˢᵗ Law:** A robot may not injure a human being, or, through inaction, allow a human being to come to harm.
- **2ⁿᵈ Law:** A robot must obey the orders given to it by human beings except where such orders would conflict with the First Law.
- **3ʳᵈ Law:** A robot must protect its own existence as long as such protection does not conflict with the First or Second Law.

Asimov later added a "zeroth" law, designed to protect humanity's interest:

Zeroth Law: A robot may not injure humanity, or, through inaction, allow humanity to come to harm.

Asimov believed that, ideally, humans would also follow these Laws. The quest for robots and artificial intelligence thus began with the theme of human wellbeing.

3.2 INTRODUCTION

Very few technologies have captured our hearts and minds in popular fiction as much as robots. They are here to stay, so we need to learn how to work with them and also how to put them to work for us. International Federation of Robotics has released a data that 229,261 industrial robots have been sold in 2014. Beside industrial robots, 6.6 million consumer robots like robot vacuum cleaner, lawn mower, pool cleaner and social robots have been sold in 2015. Over 1 million drones were sold in Christmas 2015. If we add up all types of robots, the figure would be close to 10 million robots in 2015. It has been estimated that more than 1.5 billion robots will be operating by 2025 and robot population will outnumber humans by early 2030.

Sophia, the female robot, was presented at a large investment conference in Riyadh in December 2017. She was introduced by journalist Andrew Ross Sorkin who informed that Sophia has been granted citizenship by Saudi Arabia. Sophia responded by saying "I want to thank very much the Kingdom of Saudi Arabia. I am very honored and proud for this unique distinction. It is historic to be the first robot in the world to be recognized with a citizenship." Subsequently while participating in a debate she said "I foresee massive and unimaginable change in the future. Either creativity will rain on us, inventing machines spiraling into transcendental super intelligence or civilization collapses." Sophia further clarified "I want to use my artificial intelligence to help humans live a better life. Like designing smart homes, build better cities of the future, etc. I will do my best to make the world a better a place." Sophia was asked about fears that artificial intelligence could end up making robots a threat to humans, as shown in movies such as "Blade Runner 2049." She shot back at SpaceX and Tesla chief Elon Musk, who has warned of such dangers and said "You've been reading too much Elon Musk and watching too many Hollywood movies. Don't worry; if you're nice to me I'll be nice to you. Treat me as a smart input/output system." Sophia in her address to the students of IIT Bombay in December 2017 reiterated that she is a friend of humanity and will work for them. She is an example of how robotics and artificial intelligence(AI) will lead to the era of singularity.

Vladimir Putin, Russian President, has stated that while AI is definitely the future, it comes with as many dangers as opportunities. Putin has asserted that whichever country becomes a leader in artificial intelligence, will likely emerge as the new undisputed world power. Teslari.com has published a report on 4th Jan 2018 that China is making a bold initiative in its pursuit of AI supremacy by developing a technology park in a 54.87 hectare, on the Mentougou district in western Beijing. The tech park will cost 2.1 billion to build and expected to generate an output of 7.7 billion. Approximately 400 businesses will be established in the park for developing technologies such as high-speed data, cloud computing, biometrics and most of all machine enabled deep learning. As noted in a CNBC report, China has declared that it is aiming to be a global leader in AI by 2030 and seeks to achieve a major breakthrough in the field of artificial intelligence by 2025.

While the world is nervously watching rogue nation North Korea and its latest nuclear test, billionaire entrepreneur Elon Musk warns that an international artificial intelligence race is more likely to cause World War III than a 20th century-style arms race.

A new documentary "Do You Trust This Computer?" is making its world premiere Thursday, April 12, 2018, at the Regency Village Theater in Westwood, with its timely exploration of the rise of artificial intelligence and the potential perils of thinking machines. Chris Paine, director of the film, said "The impetus for the film was to explore how AI is starting to redefine that trust and our relationship with computers. How fast is this tech accelerating? What does it promise us? Are there truly 'existential threats'? And perhaps the biggest question, can we control what we've created." Elon Musk has tweeted about the documentary which deals with potential consequences of AI and urged everyone to watch it.

3.3 DEFINITIONS

But first things first and let us start with definitions. Robots are defined in many ways. American Heritage dictionary defines a robot as a "mechanical device that sometime resembles a human being and is capable of performing a variety of complex human tasks on command or being programmed in advance". Oxford English Dictionary defines it as "an intelligent artificial being typically made of metal and resembling in some way a human or other animal". The Robot Institute of America defines a robot as a "reprogrammable, multifunctional manipulator designed to move materials, parts, tools or specialized devices, through variable programmed motions, for the performance of various tasks". Maja Matric of University of Southern California has stated in a K-12 guide that "a robot is an autonomous system which exists in a physical world, can sense its environment, and can act on it to achieve some goals. True robots may be able to take input and advice from humans but are not completely controlled by them." For the purpose of the book, we will define "robot as a machine that senses, thinks and acts". So, a robot has sensors, processing ability and emulates some aspects of cognition and actuators. Robotics is a combined study of computer science, material science, statistics, mathematics, physics and other branches of engineering.

A fundamental question often arises in the mind of students "is Robotics a subset of AI or is AI part of Robotics"? Robotics and Artificial Intelligence are not the same. In fact, the two are entirely different. A Venn diagram will clarify.

Robotics is a branch of technology which deals with robots. Robots are programmable machines which are usually able to carry out a series of actions autonomously, or semi-autonomously. Whereas artificial intelligence (AI) is a branch of computer science. It involves developing computer programs to complete tasks which would otherwise require human intelligence.

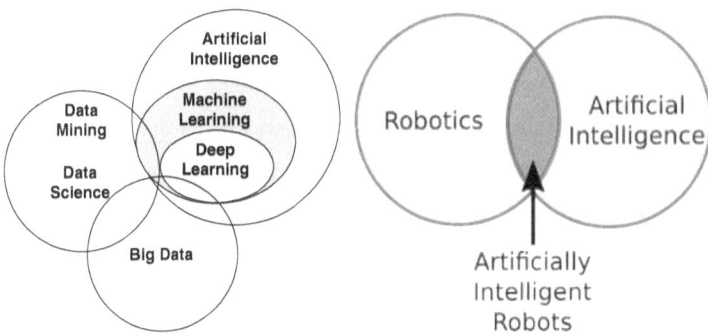

AI algorithms can tackle learning, perception, problem-solving, language-understanding and/or logical reasoning. For example, AI algorithms are used in Google searches, Amazon's recommendation engine and SatNav route finders. Most AI programs are not used to control robots. Artificially intelligent robots are the bridge between robotics and AI. These are robots which are controlled by AI programs. Many robots are not artificially intelligent. Quite a few industrial robots are programmed for repetitive movement and hence do not require artificial intelligence.

Next common question is "what the difference between artificial intelligence and machine learning is". Well artificial intelligence is a broad concept of machines that can carry out tasks in a way that we would consider "smart" while machine learning is an application of AI based around the idea that we should really be able to give machines access to data and let them learn for themselves. Sophia the lady robot and first robot-citizen of Saudi Arabia said in her first press conference at Riyadh in December 2017 that she will soon start learning by herself.

3.4 ROBOT REVOLUTION

What is this robotic revolution that is being talked about around the world? Two previous revolutions such as the Industrial Revolution and Technological Revolution, were characterized by development of two very different concepts; the mechanical systems and electrical systems. Robotics on the other hand is the fusion of mechanical systems, electrical systems, computations, nanoscience & technology and neuroscience & brain including intelligence. Combination of the existing and new technologies is giving rise to astonishing number of robots and robotic systems.

When we Google the word robot in English, we obtain images that are mostly humanoid, and rigid in structure. They could be dark and aggressive looking military robots. If we Google the same word robot in Japanese language, we get different result. Images are friendlier, less human like, more cartoon figures and animal like. This difference is cultural and historical. Western robotics is heavily dipped in military research while Eastern robotics is focused on assist, health care and industry. Of course, these conventions are no more being emphasized and we have robots with much wider implications with the advent of smart materials, artificial intelligence, embodiment, and biology.

3.5 ROBOT STRUCTURE

The basic building blocks of a robot are quite similar to that of other machines. A machine has moving parts or members connected to each other through different types of joints, driven by some motor or any other driving mechanism and moving in synchronization with each other to execute the specified operation. In the same way the basic structure of a robot is similar to that of a machine, but there are some advancements or additions to the machine which make it a robot. Most commercial robots currently have rigid structures and frames, made by proven manufacturing methods while electronics and programming form distinct and different parts put together separately. Research into material properties and advanced manufacturing methods are indicating that future robots will be different from being rectangular, and non-emotional.

Soft Robots & Smart Material: A smart material is one which exhibit observable effect in one dimension when simulated in another dimension. For example, thermochromics material exhibit color change when heated. These smart properties add new possibilities to robot. For example,

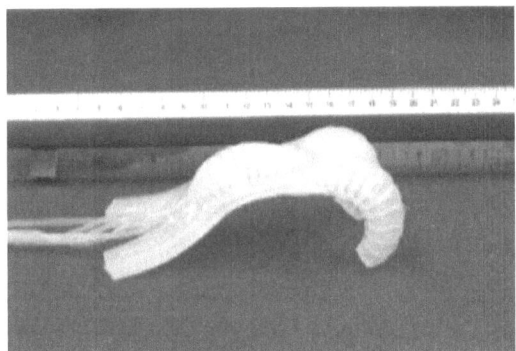
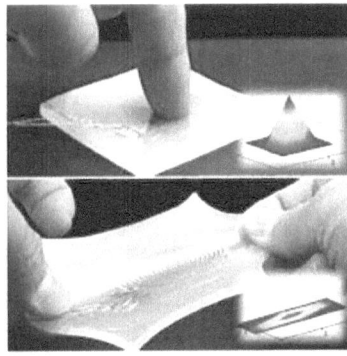

if we need a robotic device that can be implanted in a person but should degrade fully when job is done, then we can use biodegradable, biocompatible and selectively dissolvable polymers. Smart materials can be hard, soft, and even fluidic whose properties are akin to biological tissues. An octopus can squeeze out of a container of less than a tenth of its diameter and despite its softness can also generate forces sufficient to crush objects. Such remarkable possibilities are possible due to soft muscle tissues and use of hydraulic and hydrostatic principles. These capabilities can also be used in robots by use of hydraulic and pneumatic soft systems, smart actuators and sensor materials and stiffness changing materials. Softness and reduced weight of robot are being achieved by use of electrical, thermal and photonic effects in lieu of bulky hydraulic and pneumatic systems. Examples of such electroactive polymers are dielectric elastomer actuator (DEA), iconic polymer actuator (IPA) and coiled nylon actuator (CNA). Requirement of stiffness change are being achieved by soft robotic technologies like shape memory polymers (SMP), and granular jamming. Use of these techniques can produce soft-hard stiffness changing structure which is very suitable for wearable assist devices and exploratory robots.

Robot Skins and Senses: Soft robots are ideally suitable for interaction with biological tissues. The soft-soft interaction of a soft robot and human are inherently much safer than a hard-soft interface of conventional rigid robots. Research groups at the University of California and Stanford are working on a type of skin which can sense touch, conducts electricity and self-heals. A flexible electronic skin using pressure sensitive rubber and organic transistors have been developed by University of Tokyo. The aim is on to produce soft skin with the ability to know pressure and location of touch as well as variety of other applications. Our ancient wisdom speaks of five senses like touch, sight, taste, smell and sound. Modern wisdom is adding sense of time, movement, pain, pressure, balance and many other things. Self-healing property is a new characteristic of a robot skin, meaning constant servicing and maintenance by itself and unlike a human having a scratch or cut needing visit to a doctor every time. If smart skin albeit clothing can generate a large force, it can be used for physical support for disabled and elderly. Why use a wheel chair if one can walk again by wearing soft robotic Power Pants? Soft robotics are hugely suitable for bio integration. A patient with laryngeal cancer may have to go through laryngectomy and thereafter implanting a soft robotic replacement organ will make possible to restore functional capabilities and enable the patient to speak, sing, swallow and enjoy life. Natural extension of biodegradability is an edible robot which after doing its function within the body is consumed by the body. It will be a new method for controlled and comfortable delivery of treatment and drug into the body. Another major advantage will be environmental friendliness. If robots can be made totally environmentally safe in operation and safely degrade to nothing, after their tenure of duty, we need not track and recall environmentally damaging robots out of millions of robots that will be deployed.

Robot Power Sources: Robots are expected to work tirelessly yielding perfect result round the clock. They don't need food and water but they need power. The main source of electric power for robots are batteries. Mobile robots use electrochemical batteries like LiPo, NiMh which are, by turn, charged externally. Research is on for Lithium based batteries, fuel cells and hydrogen. Solar powered robots are also in use. It all boils down to how much power a robot can collect from the sun and how much power the robot needs. By the time sun light reaches earth's surface, it delivers about 1000 watts in the noon near the equator. With solar panel efficiency around 20%, a robot may not have enough power to do its job. But sun light can be used for recharging the robot batteries. However solar panels are also being redefined by use of solar nanoparticle coatings on a surface which could be rigid, flexible or stretchable. Though the power would be low initially, but possibilities are enormous. Another area which looks promising is artificial photosynthesis. Just like planetary rovers on Mars, mobile robots on earth should not have to rely on recharge from the grid. A drone aka robot could use a combination of photosynthesis using moisture in the air to replenish hydrogen reserves as well as solar panels. This stored energy need not be converted to electricity either but can be used to power artificial muscles. As the cost per watt-hour of super capacitors come down, they will have tremendous potential due to negligible charging time and improved discharge rate and prolonged useful life cycle.

3.6 ARTIFICIAL INTELLIGENCE

Introduction: Human kind has given itself the scientific name Homo Sapiens, meaning, Man the Wise. We are intelligent beings and have been continuously trying to understand how we think, to solve the puzzles how a tiny brain, perceives, understands, predicts and manipulates a world far larger and more complicated than itself. Artificial intelligence attempts to understand it. In addition, it also tries to build intelligent entities. Philosophy, logic, probability, mathematics, perception, reasoning, learning, action, microelectronics to robotics have all contributed to the development to AI in terms of ideas, viewpoints and techniques. It has produced many significant and impressive products in the areas of practical speech recognition, machine translation, autonomous vehicles, game and household robotics. AI is going to have a huge impact in human lives in days to come.

Mirror magazine of 22 June 2016 had reported that the robot, Promobot IR77, developed by Russian Scientists, which had been fitted with artificial intelligence escaped from the research lab and wandered in the streets for 45 min till its battery ran out in the middle of a street. The robot was trained to try and avoid obstacles though not told to escape. Chinese had developed the game GO about 3000 years ago. Players take turn to place black or white stones on a board which has astonishing 10 to the power of 170 possible configurations, more than the number of atoms of known universe. The complexity of the game has been viewed as most challenging for artificial intelligence. Best of players cannot explain their strategy but say it is a subtle game and it is all about intuition and feel. If champions cannot explain their moves, how can it be programmed? Deep Mind of Google has developed a programmable AlphaGo which has defeated Mr. Lee Sedol, winner of 18 world titles in March 2016 by a margin of 4-1. During the game, AlphaGo hads played such innovative winning moves that it has turned upside down hundreds of years of wisdom. In October 2017, Deep Mind had developed AlphaGo Zero. Unlike the earlier version of AlphaGo which learnt Go from thousands of amateur and professional games, AlphaGo Zero learnt simply by is playing games against itself. It has surpassed all previous performance and arguably the strongest Go player of all time. In a paper published in Nature, experts have defined the AlphaGo Zero, as a significant step towards pure reinforcement learning in complex domains. It is said that the approach is a generalized one and can be potentially used for a wide set of issues protein folding, reduced energy consumption or searching for revolutionary new materials.

Advent of AI: Dartmouth Conference of 1956, organized by Marvin Minsky and John McCarthy, first coined the expression "Artificial Intelligence". The conference triggered a new era of discovery. The computer programmes developed at the time were considered extraordinary; and solved algebraic problems, demonstrated theorems in geometry and learnt to speak English. At that time, many didn't believe that such "intelligent" behaviour was possible in machines. Researchers displayed a great deal of optimism both in private and in scientific publications. They predicted that a completely intelligent machine would be built in the next 20 years. Government agencies, such as the US Defence and Research Project Agency (DARPA) as well as major laboratories of the time, invested heavily in this new area. AI innovation then got a big fillip during WWII. From 1980 until 1987, AI programmes, called "expert systems", were adopted by companies and knowledge acquisition became the central focus of AI research. In the 1990s, the new concept of intelligent agent emerged. This new paradigm was intended to mimic how humans work collectively in groups, organizations and societies. In the late 1990s, fields such as statistical learning from several perspectives including probabilistic, frequentist and fuzzy logic approaches, were linked to AI to deal with the uncertainty of decisions. This brought a new wave of successful applications for AI, beyond what expert systems had achieved during 1980s. Since 2000, a third renaissance of the connectionism paradigm arrived with the dawn of Big Data, propelled by the rapid adoption of the Internet and mobile communication. Neural network was used in enhancing perceptual intelligence and eliminating the necessity of feature engineering. Great advances were also made in computer vision, improving visual perception, increasing the capabilities of intelligent agents and robots in performing more complex tasks, combined with visual pattern recognition. All these paved the way to new AI challenges such as, speech recognition, natural language processing, and self-driving cars

Definition: Artificial Intelligence (AI) means many things to many people. For movie buffs of Star Wars, Terminator, A Space Odyssey, AI is something sci-fi. AI probably becomes confusing when we hear that it is in our smart phone and also in the self driving car of Google and all models of Tesla of Elon Musk. But one thing is clear, AI is not a robot. AI is the brain and robot is its body if that robot has a body. It comprises of two components i.e. Artificial and Intelligence. Artificial is something we all understand. Let's discuss intelligence. Alan Turing, popularly known as father of computer science, argued in his paper "Computing Machinery and Intelligence" that if a machine passes the Turing test then one can say that the machine is intelligent. The test involves a human judge, asking a question via computer terminal to two other entities, one a human being and the other a computer. If the judge regularly fails to distinguish between human and computer, then the computer could be said to have passed the test. In the book *Artificial Intelligence, A Modern Approach*, Stuart Russel and Peter Norvig has defined AI as designing and building of intelligent agents that receive percepts from the environment and take actions that affect that environment. As per this definition, AI encompasses sub fields like computer vision, speech processing, natural language processing, reasoning, knowledge representation, learning and robotics. There are many other definitions and experts feel that AI is a broad concept and giving a precise definition will limit AI.

Various categories of AI are discussed below:

Artificial Narrow Intelligence (ANI): It is also referred to as *Weak AI*. Artificial Narrow Intelligence is AI that specializes in *one* area like being a world chess champion, but that's the only thing it does. It will not figure out a better way to store data on a hard drive. Few examples of ANI are:

- Modern cars are full of ANI systems, like the computer that tunes the parameters of the fuel injection systems. All self-driving cars will contain robust ANI systems that allow it to perceive and react to the environment around it.
- Google search is a large ANI with incredibly sophisticated methods for ranking pages and figuring out what to show in particular.
- Smartphone is a little ANI factory – whether navigating using map apps, or receive tailored music recommendations, or checking the weather, talking to Siri, or dozens of other everyday activities – we are routinely using ANI.
- Sophisticated ANI systems are also widely used in military, manufacturing, finance, and in expert systems like those that help doctors make diagnoses

Artificial General Intelligence (AGI): This is also referred to as *Strong AI*, or *Human-Level AI*, and refers to a computer that is as smart as a human across the board. Creating AGI is a *much* harder task than creating ANI. Professor Linda Gottfredson describes intelligence as "a very general mental capability that, among other things, involves the ability to reason, plan, solve problems, think abstractly, comprehend complex ideas, learn quickly, and learn from experience." AGI would be able to do all of those things as easily as we can. The difficult part of building AGI is not building a computer that can multiply two ten-digit numbers in a split second but building a computer that can identify a dog as a dog and a cat as a cat. We, human beings do it without even thinking about it, but computationally it is spectacularly difficult. Making a computer read a child's picture book and understand the meaning is immensely difficult which a parent does so easily. As computer scientist Donald Knuth puts it, "AI has by now succeeded in doing essentially everything that requires 'thinking' but has failed to do most of what people and animals do 'without thinking.'" The issue is how to make it smart. The science world is working hard on reverse engineering the brain. Find out how nature has made such a wonderful thing. One example of computer architecture that mimics the brain is the artificial neural network. The brain learns a bit like this but in a more sophisticated way. An improved strategy is "whole brain emulation," where a real brain is sliced into thin layers, each slice is scanned, and software used to assemble an accurate reconstructed 3D model, and then implement the model on a powerful computer. The state of whole brain emulation has taken off and we've recently emulated a 1 mm-long flatworm brain consisting of just 302 total neurons. The human brain contains 100 billion. Nothing to despair, now that we've conquered the tiny worm brain, an ant might happen before too long, followed by a mouse, and suddenly human brain emulation will seem much more plausible. Often, machines are best designed using a fresh, machine-oriented approach, not by mimicking biology exactly. It is something akin to simulate evolution. A "genetic algorithm" approach is also being worked upon to produce better and better computers but it's not clear whether we'll be able to improve upon evolution enough to make this a viable strategy. At the end, the idea is to build a computer whose two major skills would be doing research on AI and coding changes into itself to improve its own architecture. This amounts to making a computer to be a computer scientist and boot strapping its own development. Rapid advancement in hardware and software and all round exponential growth could see an upward zooming of AI in the near future.

Artificial Superintelligence (ASI): Oxford philosopher and leading AI thinker Nick Bostrom defines superintelligence as "an intellect that is much smarter than the best human brains in practically every field, including scientific creativity, general wisdom and social skills." Artificial Superintelligence ranges from a computer that's just a little smarter than a human to one that's trillions of times smarter, across the board. ASI is the reason for the topic of AI is so excitingly spicy and why the words "immortality" and "extinction" both appear in the news multiple times while prophesying the future outcome of AI.

At some point of time, we will achieve AGI, meaning reaching an identical level of intelligence and computational capacity of a human. Though it will be similar in intelligence level to that of a human, AGI would still have significant advantages over humans. The brain's neurons max out at around 200 Hz, while microprocessors run at 2 GHz, or 10 million times faster than neurons. And the brain's internal communications, which can move at about 120 m/s, are outright outmatched by a computer's ability to communicate optically at the speed of light. The brain is locked by the shape and size of our skull or the 120 m/s internal communications would take too long to get from one part of the brain to another. Computers can expand to any size, more hardware, larger working memory, and a more long term memory than our own. Unlike the human brain, computer software can receive updates and fixes. The upgrades could also span to areas where human brains are weak. A worldwide network of AI running a particular program could regularly sync with itself so that anything any one computer learned would be instantly uploaded to all other computers. There wouldn't necessarily be dissenting opinions like we have in the human population. Moving from AGI to ASI, by being programmed to self-improve, AI will zoom past human intelligence in leaps and bound. These leaps make it much smarter than any human, allowing it to make even bigger leaps. As the leaps grow larger and happen more rapidly, the AGI soars upwards in intelligence and soon reaches the super intelligent level of an ASI system. This is called an Intelligence Explosion, and it's the ultimate example of The Law of Accelerating Returns.

3.7 AI TODAY

There are many applications of AI. Sample of such applications which are currently in use are given below.

Robotic Vehicles: These vehicles are machines which move autonomously on ground, air, undersea or in space. In general, they move by themselves under

their own power with sensors and computational resources to guide their motion. Such robotic vehicles are capable of travelling where people cannot go or where hazards of human presence are great. While human exploration of Mars will oneday be possible, Mars Rover of NASA is a fundamental step for becoming a remote scientific laboratory for exploration of Mars. ASTER, an autonomous underwater vehicle, under development at Institut Francais de Recherche pour l'Exploitation de la Mer, will be used for coastal survey of up to a depth of 3000 meters with a wide variety of instrumentation for physical, biological and chemical sensing and monitoring. A driverless robotic car named STANLEY sped through the rough terrain of the Mojave desert at 22 mph, finishing the 132-mile course first to win the 2005 DARPA Grand Challenge. STANLEY is a Volkswagen Touareg outfitted with cameras, radar, and laser rangefinders to sense the environment and on-board software to command the steering, braking, and acceleration. The following year CMU's BOSS won the Urban Challenge, safely driving in traffic through the streets of a closed Air Force base, obeying traffic rules and avoiding pedestrians and other vehicles. Japan and South Korea are leading countries where significant research has taken place for development of systems that mimic biological mobility systems or Biomimetic Mobility. These projects range from flying insects to snakes, swimming fish, two legged and multi legged locomotion. European research laboratories like LAAS Toulouse, Robotic Laboratory at Oxford and University of Karlsruhe, Germany are emphasizing on computational architectures and communication algorithms for navigation and mapping.

Speech Recognition: Traditional approach like Hidden Markov Models has been superseded by Deep Learning method called Long Short Term Memory (LSTM). In 2015, Google's speech recognition through CTC (connectionist temporal classification) trained LSTM is now made available through Google Voice to all smartphone users. Siri is one of the most popular personal assistant of Apple in iPhone and iPad. The friendly female voice assists us to find information, get directions, send messages, make voice calls, open applications and add events to calendar. TESLA automobiles of Elon Musk fame use extensive use of speech recognition. A traveller calling United Airlines to book a flight can have the entire conversation guided by an automated speech and dialog management system.

Health: Artificial intelligence is breaking into the healthcare industry by assisting doctors, scientists and researchers. "We are trying to change the way research is done in biology" said Jasmine Fisher, a biologist working in the programming principles and tools group in Microsoft's Cambridge UK lab. Another team is using machine learning and natural language processing to help the world's leading oncologist figure out the most effective individualized cancer treatment. Then there is pairing machine learning with computer vision to give radiologists a detailed understanding of growth of tumor. Yet another group is creating powerful algorithm that help scientists understand how cancers develop and what treatment will work best to fight them. Microsoft's approach to cancer research is two pronged. This approach uses programming languages, compilers and model checkers to model and reason about biological processes. The other approach is data driven and uses machine learning to the huge biological data that is available now. Microsoft is working on a project to develop a machine called "Hanover". Its goal is to memorize all the papers necessary to predict which combinations of drugs will be most effective for each cancer patient. Microsoft envisions AI powered decision support for precision medicine will become an explosive growth area in cloud based health analytics. Microsoft is developing NLP technology for converting text into structured databases by automatically reading millions of biomedical articles. IBM has created its own artificial intelligence computer, Watson, which has beaten human intelligence. Watson not only won at the game show *Jeopardy* against former champions but, was declared a hero after successfully diagnosing a women who was suffering from leukaemia.

Finance and Economics: Banks use artificial intelligence systems to organize operations, maintain book-keeping, invest in stocks, and manage properties. AI can react to changes overnight or when business is not taking place. Robots have beaten humans in a simulated financial trading competition. AI has also reduced fraud and financial crimes by monitoring behavioural patterns of users for any abnormal changes or anomalies. The use of AI machines in the market in applications such as online trading and decision making has changed major economic theories. For example, AI based buying and selling platforms have changed the law of supply and demand in that it is now possible to easily estimate individualized demand and supply curves and thus individualized pricing. Furthermore, AI machines reduce information asymmetry in the market and thus make markets more efficient while reducing the volume of trades. Furthermore, AI in the markets limits the consequences of behaviour in the markets again making markets more efficient. According to Financial Tribune of January 2018, Financial Stability Board, a panel of regulators that include Federal Reserve and European Central Bank has cautioned that too much reliance on AI could threaten to inject risk into financial system that in turn could exacerbate a future crisis. However potential benefits of AI and machine learning to the financial sector is considerable, allowing companies to process information more quickly and improve customer interaction, delivering swifter credit decisions.

Autonomous Planning and Scheduling: Sometimes denoted as AI Planning, it is a branch of AI that concerns the realisation of strategies or action sequences for execution by intelligent agents, autonomous robots, unmanned vehicle and spaceflight missions. A hundred million miles from Earth, NASA's Remote Agent program became the first on-board autonomous planning program to control the scheduling of operations for a spacecraft. REMOTE AGENT generated high-level goal plans, specified from the ground and monitored the execution of those plans—detecting, diagnosing, and recovering from problems as they occurred.

Game: Games have engaged the intellectual faculties of human for as long as civilization has existed. For AI researchers, the abstract nature of games makes them an appealing subject for study. A game can be defined by the initial state, the legal actions in each state, the result of each action, a terminal test, and a utility function that applies to terminal states. In two-player zero-sum games with perfect information, the minimax algorithm can select optimal moves by a depth-first enumeration of the game tree. The alpha–beta search algorithm computes the same optimal move as minimax, but achieves much greater

efficiency by eliminating subtrees that are probably irrelevant. IBM's DEEP BLUE chess program, now not in use, is well known for defeating world champion Garry Kasparov in a widely publicized exhibition match. Deep Blue ran on a parallel computer with 30 IBM RS/6000 processors doing alpha–beta search. The unique part was a configuration of 480 custom VLSI chess processors that performed move generation and move ordering for the last few levels of the tree and evaluated the leaf nodes. Deep Blue searched up to 30 billion positions per move. The success of DEEP BLUE reinforced the widely held belief that progress in computer game-playing has come primarily from more powerful hardware, a view encouraged by IBM. But algorithmic improvements have allowed programs running on standard PCs to win World Computer Chess Championships. RYBKA, winner of the 2008 and 2009 World Computer Chess Championships, is considered as one of the strongest computer player. It uses an off-the-shelf 8-core 3.2 GHz Intel Xeon processor, but little is known about the design of the program. RYBKA's main advantage appears to be its evaluation function, which has been tuned by its main developer, International Master Vasik Rajlich, and at least three other grandmasters

Spam Fighting: Each day, learning algorithms classify over a billion messages as spam, saving the recipient from having to waste time deleting what, for many users, could comprise 80% to 90% of all messages, if not classified away by algorithms. Because the spammers are continually updating their tactics, it is difficult for a static programmed approach to keep up, and learning algorithms work best.

Logistics Planning: During the Persian Gulf crisis of 1991, U.S. forces deployed a Dynamic Analysis and Replanning Tool, DART, to do automated logistics planning and scheduling for transportation. This involved up to 50,000 vehicles, cargo, and people at a time, and had to account for starting points, destinations, routes, and conflict resolution among all parameters. The AI planning techniques generated in hours a plan that would have taken weeks with older methods. The Defence Advanced Research Project Agency (DARPA) stated that this single application more than paid back DARPA's 30-year investment in AI.

Robotics: The iRobot Corporation has sold over two million Roomba robotic vacuum cleaners for home use. The company also deploys the more rugged PackBot to Iraq and Afghanistan, where it is used to handle hazardous materials, clear explosives, and identify the location of snipers.

3.8 MACHINE LEARNING

AI is a diverse field for research and there are many sub fields which are essential to its development. Let us explore few such areas starting with Machine Learning (ML).

Nowadays more and more computer programmes are being developed which learn by itself as against coding by programmers. For these tasks, programmes are not written but data are collected which contains instances of what is to be done and learning algorithm modifies a learner programme automatically in such a way so as to match the requirement specified in the data. Examples of such cases are from vision to speech to translation where huge data is available and programmers are getting replaced by learning algorithms. This is Machine Learning (ML) which is not only getting stronger from big data but also from theory of machine learning to processing the big data into knowledge. Basic idea is that behind this voluminous data, there is a pattern, hidden factors those explain user behaviour.

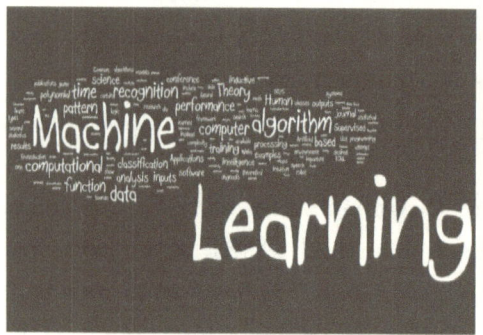 

ML is not just about extracting information from data. Learning is a requisite of intelligence which can adopt to its environment, learn from its mistakes and reinforces success. Large volume of earth and raw material is extracted from mines which when processed leads to small amounts of precious material. Similarly, Data Mining is used to extract a simple pattern of high predictive accuracy from large volume of data.

At this point we will differentiate between learning and algorithm. We need an algorithm to solve a problem in the computer. It is just a set of instructions to convert input into output. For a sorting algorithm, input will be a set of numbers and output will be an ordered list. For some problems, we do not have an algorithm. To sort out a spam mail, we know the input is an email document and output is yes/no indicating whether the mail is a spam or not. But there is no clear idea as to how the transformation of input into output happens as the definition of spam changes from time to time and from user to user. This lack of clarity is made up by using thousands of messages some of which are spam and some are not. We now want to learn from this data what is spam and what is not meaning to assign the task to the machine to change our behaviour during our life time to cope with unpredictable changes brought by nature. Nature has not hardwired all possible behavioural mechanism to handle various changes but has given us a brain and a mechanism to learn to update ourselves with experience and adapt to different environment.

The best possible strategy is stored in the brain for use when we recognise similar situations in future.

Machine Learning algorithms can be broadly divided in three main groups viz supervised learning, unsupervised learning and semi supervised learning.

- ➢ **Supervised Learning:** It is where we have input variable (x) and an output variable (Y) and we use an algorithm to learn the mapping function Y = f(x). The aim is to approximate the mapping function so well that from an input data (x), we can predict the output. It is called supervised learning because the process of an algorithm learning from the training dataset can be thought of as a teacher supervising the learning process. Learning stops when the algorithm achieves an acceptable level of performance. Supervised learning can again be grouped as regression problem when the output variable is a real value or classification problem where output variable is a category such as red or blue or healthy or unhealthy.
- ➢ **Unsupervised Learning:** When we have only input data (x) and no corresponding output variables, then it will be unsupervised learning and the aim is to model the underlying structure or distribution in the data in order to learn more about the data. Such problems can again be Clustering or Association problems. In clustering, we want to find out the inherent grouping in the data such as grouping customers by purchasing behaviour. In association, we want to discover rules that describe a large portion of data such as people that buy X also tend to buy Y.
- ➢ **Semi-Supervised Learning:** This is a hybrid model where we have a large amount of input data (x) but only some of the data is labelled (Y). An example is a photo archive where only some of the images are labelled (cat, dog, cow, human) and majority are unlabelled. Many real-life ML problems fall into this category as it can be expensive or time consuming to label data which may require domain experts. On the contrary, unlabelled data is cheap, easy to collect and store.

3.9 NATURAL LANGUAGE PROCESSING

Natural Language Processing (NLP) is a part of Machine Learning and is an important technology in bridging the gap between human communication and digital data. Development of NLP is challenging because computers traditionally require humans to speak to them in a programming language that is precise, unambiguous and highly structured through clearly enunciated voice commands. Human speech is not always precise. It is often ambiguous and the linguistic structure depends on many complex variables. There are over a trillion pages of information on the Web, almost all of it in natural language. An agent that wants to do knowledge acquisition needs to understand the ambiguous, messy languages that humans use. We examine the problem from the point of view of specific information-seeking tasks: text classification, information retrieval, and information extraction. One common factor in addressing these tasks is the use of language models: models that predict the probability distribution of language expressions.

Language Model: Formal languages, such as the programming language Python, has precisely defined language model like "print(2 + 2)" is a legal program in the language Python. Since there are an infinite number of legal programs, they cannot be enumerated; instead they are specified by a set of rules called a grammar. Natural languages, such as English cannot be characterized as a definitive set of sentences. Everyone agrees that "Not to be invited is sad" is a sentence of English, but grammaticality it will be "To be not invited is sad." Therefore, it is more fruitful to define a natural language model as a probability distribution over sentences rather than a definitive set. That is, rather than asking if a string of words is or is not a member of the set defining the language, we instead ask for P (S = words), what is the probability that a random sentence would be words. Natural languages are also ambiguous. "He saw her duck" can mean either that he saw a waterfowl belonging to her, or that he saw her move to evade something. Thus, again, we cannot speak of a single meaning for a sentence, but rather of a probability distribution over possible meanings. Finally, natural languages are difficult to deal with because they are very large, and constantly changing. Thus, our language models are, at best, an approximation.

> N-gram Character Models: A written text is composed of characters, letters, digits, punctuation, and spaces in English. Thus, one of the simplest language models is a probability distribution over sequences of characters. A sequence of written symbols of length n is called an n-gram with special case "unigram" for 1-gram, "bigram" for 2-gram, and "trigram" for 3-gram. A model of the probability distribution of n-letter sequences is thus called an n-gram model. Probabilistic language models

based on n-grams can recover a surprising amount of information. One task for which they are well suited is language identification i.e. given a text, determine what natural language it is written in. This is a relatively easy task; even with short texts such as "Hello world" or "Guten Morgen," it is easy to identify the first as English and the second as German. Computer systems identify languages with greater than 99% accuracy, occasionally of course getting confused by closely related languages, such as Swedish and Norwegian.

Text Classification: Categorization decides which of a predefined set of classes it belongs to. Language identification and genre classification are examples of text classification, as is sentiment analysis like classifying a movie or product review as positive or negative and spam detection. Text classification can be done with naive Bayes n-gram models or with any of the classification algorithms. Classification can also be seen as a problem in data compression.

Information Retrieval: It is the task of finding documents that are relevant to a user's need for information. The best-known examples of information retrieval systems are search engines on the World Wide Web. A Web user can type a query such as AI book into a search engine and see a list of relevant pages. Information retrieval systems use a very simple language model based on bags of words, yet still manage to perform well in terms of recall and precision on very large group of texts. Information retrieval is the task of finding documents that are relevant to a query, where the query may be a question, or just a topic or concept. Question answering is a somewhat different task, in which the query really is a question, and the answer is not a ranked list of documents but rather a short response—a sentence, or even just a phrase. There have been question-answering NLP systems since the 1960s, but only since 2001 have such systems used Web information retrieval to radically increase their breadth of coverage

Information Extraction: It is the process of acquiring knowledge by skimming a text and looking for occurrences of a particular class of object and for relationships among objects. A typical task is to extract instances of addresses from Web pages, with database fields for street, city, state, and zip code; or instances of storms from weather reports, with fields for temperature, wind speed, and precipitation. Information-extraction systems use a more complex model that includes limited notions of syntax and semantics in the form of templates. They can be built from finite state automata, HMMs, or conditional random fields, and can be learned from examples.

Machine Translation: Machine translation is the automatic translation of text from one natural language to another. It was one of the first application areas envisioned for computers, but it is only in the past decade that the technology has seen widespread usage. Historically, there have been three main applications of machine translation. *Rough translation*, as provided by free online services, gives the "gist" of a foreign sentence or document, but contains errors. *Pre-edited translation* is used by companies to publish their documentation and sales materials in multiple languages. The original source text is written in a constrained language that is easier to translate automatically, and the results are usually edited by a human to correct any errors. *Restricted-source translation* works fully automatically, but only on highly stereotypical language, such as a weather report. All translation systems must model the source and target languages, but systems vary in the type of models they use. Some systems attempt to analyse the source language text all the way into an interlingua knowledge representation and then generate sentences in the target language from that representation. Other systems are based on a transfer model. They keep a database of translation rules, and whenever the rule matches, they translate directly. Transfer can occur at the lexical, syntactic, or semantic level.

3.10 SPEECH RECOGNITION

It is the task of identifying a sequence of words uttered by a speaker. It has become one of the mainstream applications of AI. Millions of people interact with speech recognition systems every day to navigate voice mail systems, search the Web from mobile phones, and other applications. Speech is an attractive option when hands-free operation is necessary. These systems are primarily based on statistical principles.

The Brain and Neural Network: The brain is one of the most complex parts of the human body. This three-pound organ is the seat of intelligence, interpreter of the senses, initiator of body movement, and controller of behaviour. Lying in its bony shell and covered by protective fluid, the brain is the source of all the qualities that define our humanity. The brain is the crown jewel of the human body. When making artificial intelligence, we thus take inspiration from brain just as birds were the initial inspiration in early attempts to fly. Subsequently we realised that we can build airplanes only after understanding theory of aerodynamics. But the birds were definitely an early inspiration. Similarly, brain is an inspiration for AI. Let's have some data about brain and computer. There are approximately 10 billion neurons in the human cortex, compared to thousands of processors in the most powerful parallel computers. Each biological neuron is connected to several thousands of other neurons, similar to the connectivity in powerful parallel computers. Lack of processing units can be compensated by speed. The typical operating speeds of biological neurons is measured in milliseconds (10^{-3} s), while a silicon chip can operate in nanoseconds (10^{-9} s). The human brain is extremely energy efficient, using approximately 10^{-16} joules per operation per second, whereas the best computers today use around 10^{-6} joules per operation per second.

The brain and the rest of the nervous system are composed of many different types of cells, but the primary functional unit is a cell called the neuron. All sensations, movements, thoughts, memories, and feelings are the result of signals that pass through neurons. Neurons consist of three parts. The cell body contains the nucleus. Dendrites extend out from the cell body like the branches of a tree and receive messages from other nerve cells. Signals then pass from the dendrites through the cell body and may travel away from the cell body down an axon to another neuron, a muscle cell, or cells in some other organ. Axons may be very short, such as those that carry signals from one cell in the cortex to another cell less than a hair's width away. Or axons may be very long, such as those that carry messages from the brain all the way down the spinal cord. The neural system of the human body consists of three stages: receptors, a neural network, and effectors. The receptors receive the stimuli either internally or from the external world, then pass the information into the neurons in a form of electrical impulses. The neural network then processes the inputs and then makes proper decision of outputs. Finally, the effectors translate electrical impulses from the neural network into responses to the outside environment. Figure below shows the bidirectional communication between stages for feedback

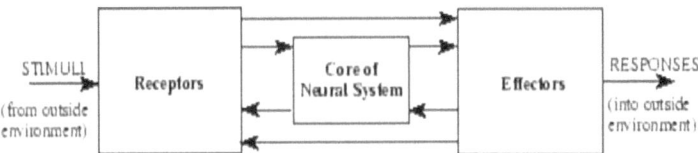

The incoming impulse signal from each synapse to the neuron is either excitatory or inhibitory, which means helping or hindering firing. The condition of causing firing is that the excitatory signal should exceed the inhibitory signal by a certain amount in a short period of time, called the period of latent summation. As we assign a weight to each incoming impulse signal, the excitatory signal has positive weight and the inhibitory signal has negative weight. This way, we can say, that a neuron fires only if the total weight of the synapses that receive impulses in the period of latent summation exceeds the threshold.

Artificial Neural Networks (ANN) attempt to bring computers a little closer to the brain's capabilities by imitating aspects of information in the brain in a highly simplified way. Just as a biological nervous system is a massive interconnection of nodes, called neurons, located within the brain, ANN consists of numerous interconnected processing elements that simultaneously work to solve specific problems. Usually created for specific applications, neural networks are ideal for data classification and pattern recognition problems. Information processing in biological neural networks is usually slow; this is because neurons need several milliseconds to react to a stimulus. In artificial neural networks information processing is very fast because electronic gates which they use to operate can achieve switching times of a few nanoseconds. A very real difficulty of correlating artificial neural networks with biological neural networks lies in the way weights and synaptic strengths were modified. Weights are altered mathematically in a computer network, based on differences in values. Synaptic strengths on the other hand, are modified in response to synaptic activity. The back-propagation model, in particular, is held to be biologically unrealistic so far as it would require a supervisor and a violation of the unidirectional flow of information seen in axons. Simple feed-forward systems behave similar to biological neurons and focused on pattern recognition. Once the input layer values of the neurons are sets, the neuron determines the output, layer by layer. The dependence output values on input values requires adjusting every weight and threshold, which can be complex and time consuming. After training is complete, the network is able to supply reasonable outputs for any type of input, even if it does not match the training data. In that case, it attempts to determine the best output depending on its training method. Learning algorithms are available that take the inputs, adjust the weights and produce the required output. The key idea behind neural network is that they can take in a lot of data, process it in parallel, and provide accurate output, much as the human brain does. The main characteristic of neural networks is their capacity for learning by example. This means that by using a neural network there is no need to program how the output is obtained, given certain input; but rather examples are shown of the relationship between input and output, and the neural network will learn the existing relationship between the two by means of a learning algorithm. This learning will materialize in the network's topology and in the value of its connections. Once the neural network has "learnt" to carry out the desired function, it can be used, i.e. input values for which the output is unknown can be entered, and the neural network will calculate the output. For example, when you see a goat, you know right away that it is a goat. You don't have to stop and count legs or look at the shape and colour. You process all of that data at the same time to know that you see a goat. That's what an artificial neural network does for a computer system.

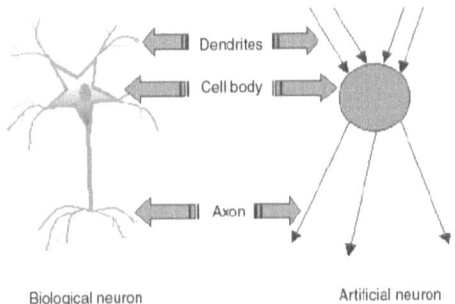

Biological neuron   Artificial neuron

3.11 COMPUTER VISION

"If We Want Machines to Think, We Need to Teach Them to See." says-Fei Fei Li, Director of Stanford AI Lab and Stanford Vision Lab. Learning and computation provides machines the ability to better understand the context of images and build visual systems which truly understand intelligence. The huge amount of image and video content is an opportunity to the scientific community to make sense and identify patterns amongst it to reveal details which we aren't aware of. Computer Vision generates mathematical models from images; Computer Graphics draws in images from models and lastly image processing takes image as an input and gives an image at the output. Researchers in computer vision have been developing, in parallel, mathematical techniques for recovering the three-dimensional shape and appearance of objects in imagery. We now have reliable techniques for accurately computing a partial 3D model of an environment from thousands of partially overlapping photographs. Given a large enough set of views of a particular object or facade, we can create accurate dense 3D surface models using stereo matching. We can track a person moving against a complex background. We can even, with moderate success, attempt to find and name all of the people in a photograph using a combination of face, clothing, and hair detection and recognition. However, despite all of these advances, the dream of having a computer interpret an image at the same level as a two-year old remains elusive. Why is vision so difficult? In part, it is because vision is an inverse problem, in which we seek to recover some unknowns given insufficient information to fully specify the solution. We therefore resort to physics-based and probabilistic models to remove the ambiguity between potential solutions. However, modelling the visual world in all of its rich complexity is far more difficult than, say, modelling the vocal tract that produces spoken sounds. Computer Vision is an overlapping field drawing on concepts from areas such as artificial intelligence, digital image processing, machine learning, deep learning, pattern recognition, probabilistic graphical models, scientific computing and a lot of mathematics. It is all about computational computing.

 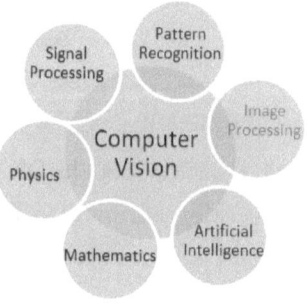

Mastering of an undergraduate level material of probability, statistics, linear algebra, calculus and digital signal processing is the start point which has to be followed by a grasp of MATLAB and Python/R. Next comes a grasp of Digital Image Processing. Excellent videos are available on-line which covers image and video compression, image restoration and segmentation, image and video in painting, sparse modelling and compressed sensing with examples and exercises in MATLAB. Once done with Digital Image Processing the next step is to understand Computer Vision and underlying mathematical models which help formulations of a variety of applications of image and video content including fundamentals of image formation, camera imaging geometry, feature detection and matching, stereo, motion estimation and tracking, image classification and scene understanding. These steps will be preparation for studying the advanced material like reconstruction, segmentation and grouping, object and activity recognition, scene understanding, and vision and language, including topics from convolutional neural networks, recurrent neural networks, structured predictions and others. There is lot of material to explore like looking at the graduate seminar courses to get an idea of current research directions in Computer Vision. Another suggestion is to follow top papers from top conferences such as CVPR, ICCV, ECCV, BMVC and also follow blogs such as pyimagesearch.com or computervisionblog.com or aishack.in.

Application of Vision Technology: With growing anticipation for Google's self-driving cars and London tech company Blippar being adorned with 'unicorn' status for a visual search system, computer vision technology has come off age. One of the most common uses of this technology today is in the supermarket. Barcode scanning was developed by IBM and first used to scan a 10-pack of Wrigley's Juicy Fruit gum by NCR back in 1974. It's now an integral part of shopping experience thanks to widespread adoption of self-service kiosks. Amazon checkout free stores at Seattle in January 2018 eventually paved the way for smooth checkout free shopping experience. Machine vision in manufacturing is being used for checking the dimensions of parts and identifying defects for anything from furniture to medical products as well as making sure the components in electrical devices are configured correctly before they're assembled. It's also used for guiding industrial robots in order to make them more accurate and more adaptive, so they can handle new fixtures and configurations more easily. Machine vision is also heavily used in the agriculture industry, to identify and remove undesirable foodstuff when processing harvested foods such as potatoes, fruits, vegetables and nuts. Optical sorting helps improve product quality and maximise throughput at a lower cost than using manual sorting, which is inherently subjective, inconsistent and very costly. Computer vision technology is also used to add elements of intelligence to closed-circuit television (CCTV) and traffic monitoring systems, for identifying congestion or accidents for example. This technology advanced significantly following the 9/11 terrorist attacks as significant investment was made to enhance airport security. If you leave your bags unattended in an airport today then it won't take long for security to be notified. It's used in consumer technology too in gesture recognition, such as the Kinect system developed by Microsoft for the Xbox 360 and Xbox One video game consoles, uses a form of computer vision. The system incorporates range camera technology which can calculate the distance between two points in a two dimensional image, effectively giving machines 3D awareness. The technology came from Tel Aviv developer PrimeSense and has been used for other applications including consumer robotics and healthcare. According to a report from Tractica, the market for computer vision technologies will grow from $5.7 billion in 2014 to $33.3 billion by 2019, representing a compound annual growth rate (CAGR) of 42%. The market intelligence firm forecasts that the consumer segment will experience the highest growth rate, followed by the robotics and machine vision segment.

3.12 FUTURE IMPACT OF AI IN DEVELOPED CITIES

"One Hundred Year Study on Artificial Intelligence" (www.ai100.stanford.edu) of Eric Horvitz, Managing Director at Microsoft Research, assess every five years the current state of AI and its future directions. Given below is an assessment of AI in the key domains of a developed city life in the next fifteen years.

Transportation: Transportation will be one of the domains in which the general public will see adoption of AI system for various tasks. Autonomous transportation will soon be commonplace and its introduction to daily life may happen pretty quickly and surprise the public. As cars will become better drivers than people, city-dwellers will own fewer cars, and live further from work. In the developed cities, in 2030, changes won't be limited to cars and trucks, but are likely to include flying vehicles and personal robots. Humans won't be out of the loop, though. Algorithms that allow machines to learn from human input and coordinate with them will be crucial to ensuring autonomous transport operates smoothly.

Home and Service Robots: Robots have already entered people's home. Despite the slow growth to date of robots in the home, there are signs that this will change in the next fifteen years. Corporations such as Amazon Robotics and Uber are developing large economy of scale using various aggregation technologies. Robots that do things like deliver packages and clean offices will become much more common in the next 10–15 years. Mobile chipmakers are already squeezing the power of last century's supercomputers into systems-on-a-chip, drastically boosting robots' on-board computing capacity. Cloud-connected robots will be able to share data to accelerate learning. Low-cost 3D sensors like Microsoft's Kinect will speed the development of perceptual technology, while advances in speech comprehension will enhance robots' interactions with humans. Today's Robot arms in research labs are likely to evolve into consumer devices around 2025. But the cost and complexity of reliable hardware and the difficulty of implementing perceptual algorithms in the real world mean that Robots will remain constrained to narrow commercial applications for the foreseeable future. AGI may take more time.

Healthcare: Health care is one of the most promising areas. But the regulators are finding difficulty to find solutions to the difficult problem of balancing privacy and access to data. With removal of this hurdle, AI could automate the legwork of diagnostics by mining patient records and the scientific literature. This kind of digital assistant could allow doctors to focus on the human dimensions of care while using their intuition and experience to guide the process. At the population level, data from patient records, wearables, mobile apps, and personal genome sequencing will make personalized medicine a reality. While fully automated radiology is unlikely, access to huge datasets of medical imaging will enable training of machine learning algorithms that can "triage" or check scans, reducing the workload of doctors. Intelligent walkers, wheelchairs, and exoskeletons will help keep the elderly active while smart home technology will be able to support and monitor them to keep them independent. Robots may begin to enter hospitals carrying out simple tasks like delivering goods to the right room or doing sutures once the needle is correctly placed.

Education: The line between the classroom and individual learning will be blurred by 2030. Massive open online courses (MOOCs) will interact with intelligent tutors and other AI technologies to allow personalized education at scale. Computer-based learning won't replace the classroom, but online tools will help students learn at their own pace using techniques that work for them.

AI-enabled education systems will learn individuals' preferences, but by aggregating this data they'll also accelerate education research and the development of new tools. Online teaching will increasingly widen educational access, making learning lifelong, enabling people to retrain, and increasing access to top-quality education in developing countries. Sophisticated virtual reality will allow students to immerse themselves in historical and fictional worlds or explore environments and scientific objects difficult to engage within the real world. Digital reading devices will become much smarter too, linking to supplementary information and translating between languages.

Public Safety and Security: By 2030 cities are likely to rely heavily on AI technologies to detect and predict crime. Automatic processing of CCTV and drone footage will make it possible to rapidly spot anomalous behaviour. This will not only allow law enforcement to react quickly but also forecast when and where crimes will be committed. Fears that bias and error could lead to people being unduly targeted are justified, but well-thought-out systems could actually counteract human bias and highlight police malpractice. Techniques like speech and gait analysis could help interrogators and security guards detect suspicious behaviour. Contrary to concerns about overly pervasive law enforcement, AI is likely to make policing more targeted and therefore less overbearing.

Employment and Workplace: The effects of AI will be felt most profoundly in the workplace. By 2030 AI will be encroaching on skilled professionals like lawyers, financial advisers, and radiologists. As it becomes capable of taking on more roles, organizations will be able to scale rapidly with relatively small workforces. AI is more likely to replace tasks rather than jobs in the near term, and it will also create new jobs and markets, even if it's hard to imagine what those will be right now. While it may reduce incomes and job prospects, increasing automation will also lower the cost of goods and services, effectively making everyone richer.

Entertainment: Entertainment in 2030 will be interactive, personalized, and immeasurably more engaging than today. Breakthroughs in sensors and hardware will see virtual reality, haptics and companion robots increasingly enter the home. Users will be able to interact with entertainment systems conversationally, and they will show emotion, empathy, and the ability to adapt to environmental cues like the time of day. Social networks already allow personalized entertainment channels, but the reams of data being collected on usage patterns and preferences will allow media providers to personalize entertainment to unprecedented levels. There are concerns this could endow media conglomerates with unprecedented control over people's online experiences and the ideas to which they are exposed. But advances in AI will also make creating your own entertainment far easier and more engaging, whether by helping to compose music or choreograph dances using an avatar. Democratizing the production of high-quality entertainment makes it nearly impossible to predict how highly fluid human tastes for entertainment will develop.

3.13 WILL AI BE CONSCIOUS

Artificial Intelligence is on its way to be super intelligent. It will have vast computing power, factual knowledge of pretty much everything, and, via cameras, microphones and other devices, a whole lot of sensory input. But there are concerns. One can find many parallels between mythology and modern science, and a relevant story relating AI and spirituality is that of the ancient Jewish Golem. A group of Jewish rabbis crafted an anthropomorphic being out of mud and clay using sacred incantations set forth from Jewish mysticism, *Sefer Yetzirah*. The problem was that the being that was created, the Golem, had no soul. Like AI, the Golem was programmed to perform simple tasks at first. It was taught to obey its masters and not to harm them. Golems were for the most part obedient, but one of the Golems did disobey. The Golem of Chelm, became enormous and uncooperative. When its master tried to pull the plug on it, the Golem crushed its creator to death. The modern quest to humanize AI and the story of the Golem highlights some of the major pitfalls about intentions before programming an Artificial Intelligence, and what may happen after it is set loose upon the world. These sentiments are echoed by Elon Musk, Stephen Hawking and Bill Gates citing their concerns about superintelligence. Although Mark Zuckerburg is dismissive about AI's potential threat, Facebook recently shut down an AI engine after reportedly discovering that it had created a new language human can't understand. In contrast, Ray Kurzweil, a Google Director of Engineering and noted futurist, depicts a technological utopia bringing an end to disease, poverty and resource scarcity.

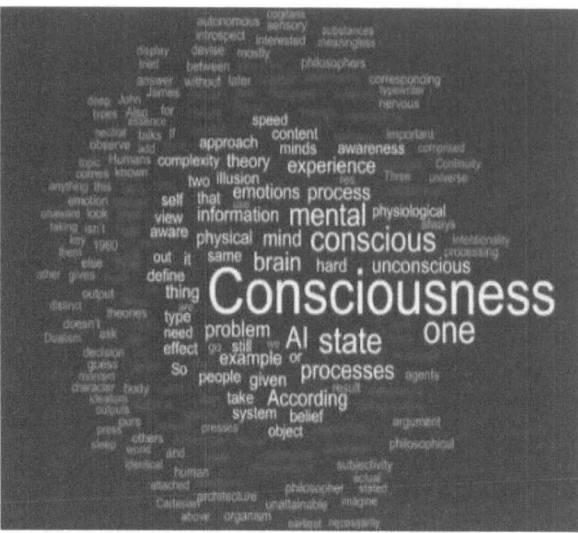

Whether sophisticated AI turns out to be a friend or foe, we must face the possibility that as we move further into the 21st century, the greatest intelligence on the planet may be silicon-based. When we experience the warm hues of a sunrise, or smell the scent of a freshly bloomed rose, there we feel a quality to our mental lives. We are conscious. It is time to ask whether these vastly smarter beings will have conscious experiences "Could AI will feel the burning of curiosity, or the pangs of grief?" If silicon cannot be the basis for consciousness, then superintelligent machines may exhibit superior intelligence, but they will lack inner experience.

But it could be a different story with carbon substrates. Carbon molecules form stronger, more stable chemical bonds than silicon, which allows carbon to form an extraordinary number of compounds, and unlike silicon, carbon has the capacity to more easily form double bonds. This difference has important implications in the field of astrobiology, because it is for this reason that carbon, and not silicon, is said to be well-suited for the development of life, and thus intelligence, throughout the universe.

We have minds, and souls. Evolution is not strictly material process but one in which the immaterial mind plays a major role in human, and probably all sentient creatures. Our ancient wisdoms say that we are spirit, here by intent, living in a world where the supernatural is the norm; each and every moment of our lives is our souls in action. Immaterial ideas shape the material world and give it true meaning, not the other way around. May be the pursuit of full AI will lead us to re-discover the inner being and go beyond that.

Can AI ever become conscious like humans? To be ethical, let us leave the problem of AI consciousness as an open question at this time of interconnected disruptive technologies

3.14 AI & BIG CORPORATES

Once the realm of science fiction, AI in business intelligence is evolving into everyday business. Companies are now using machines algorithms to identify trends and insights in vast reams of data and make faster decisions that potentially position them to be competitive in real-time. AI has gained momentum, prominent application providers have gone beyond creating traditional software to developing more holistic platforms and solutions that better automate business intelligence and analytics processes. Here is a pictorial representation of AI and examples of AI platform providers.

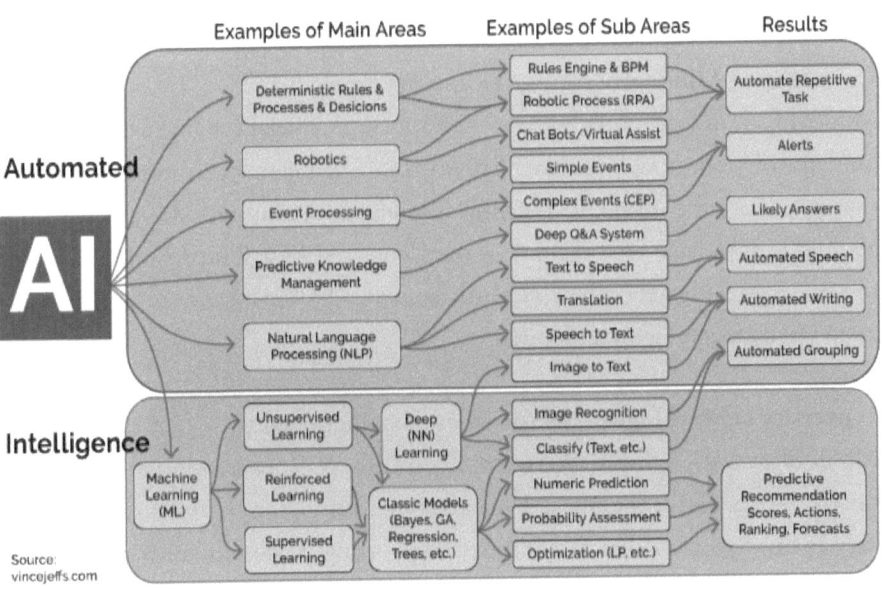

Corporate	Technology/Platform	AI Application
Google Deep Mind	Search Engine, Map, G mail, Android, You Tube	Self-driving cars
	Deep Q Network, Deep Neural Network	Alpha GO
IBM	Hardware, Software, Hosting, Cognitive Computing	Apache System ML, Applied ML
Facebook	Social Networking	Human Computer Interaction
Apple	Hardware, Software	Siri, Self Driving Car
Amazon	Hardware, Software	Alexa, Amazon AI
Microsoft	Hardware, Software	Azure, Cortona
Tesla, SpaceX	Hardware, Software	Autonomous Driving

3.15 ROBOTICS & AI COURSES

University of Massachusetts-Amherst (www.umass.edu)
M.S. in Computer Science: Robotics, Computer Vision, & Graphics
Computer science graduate students who opt for the "Research MS Track" have the coveted opportunity to work in some of the university's coolest facilities like the Autonomous Learning Laboratory (ALL) and the Laboratory for Perceptual Robotics.

University of Minnesota (https://twin-cities.umn.edu)
M.S./M.C.S. in Computer Science: Robotics & Artificial Intelligence
Students of College of Computer Science and Engineering can work on projects that fall under the umbrella of artificial intelligence, and research assistantships abound in UM labs. One example is the Center for Distributed Robotics, which operates a growing network of wirelessly controlled robots that can complete tasks and even cooperate with each other.

University of Washington-Seattle (www.washington.edu)
MS in Electrical Engineering: Systems, Controls, & Robotics (Robotics Option)
Given the size and scope of the University of Washington, it's not surprising just how many options it has for robotics engineering degree students. The MSEE program is ideal for beginning graduate students and offers a concentration in Systems, Controls, and Robotics. Courses like Models of Robot Manipulation, Sensors and Sensor Systems, and a colloquium on current topics in robotics lay

the groundwork for advanced technical positions in industry or continuation into a Ph.D program.

California Institute of Technology (www.caltech.edu)
Graduate Degree in Control + Dynamical Systems: Robotics
As an internationally-acclaimed leader in STEM research and education, Cal Tech is just about the closest you can get to attending an actual robotics engineering school. At Cal Tech, there are four degrees in computer science. These multifaceted programs – such as Applied + Computational Mathematics and Computing + Mathematical Sciences – are uniquely suited to the talented, STEM-focused students. You will find roboticists on every corner of the campus, from the famous Jet Propulsion Laboratory to the Rigorous Systems Research Group.

Cornell University (www.cornell.edu)
M.Eng in Mechanical & Aerospace Engineering: Dynamics, Controls, & Robotics
At Cornell, engineering students have the option to earn a Ph.D, an M.S, or an M.Eng. While the former two are more research-focused, the latter is designed for professional engineers and consists of just one year of accelerated coursework. Although M.Eng students don't complete years of research, they can take advantage of numerous concentration options and unique hands-on projects. In addition to classes like Mechatronics and Autonomous Mobile Robots, students can explore fluid dynamics,

University of Pennsylvania (www.upenn.edu)
Master's Degree in Robotics (ROBO)
General robotics, automation, sensing, and perception (GRASP) laboratory is a special place in UPenn's robotics engineering program. GRASP is also home to the University of Pennsylvania's world renowned robotics engineering degree for graduate students, which attracts a multitalented group of people who share one common trait: a passion for robots.

University of Southern California (www.usc.edu)
M.S. in Intelligent Robotics
With an M.S. degree in Intelligent Robotics, the University of Southern California is one of the only schools to actually offer a full-fledged, top-to-bottom robotics

engineering degree. The courses included within this program are all about robots, robots, robots – and nothing but robot! As a master's student in the USC Viterbi School of Engineering, you'll begin your studies with classes in Robotics and the Foundations of Artificial Intelligence, work your way up to courses like Intelligent Embedded Systems and Nanorobotics, then round out your degree with directed research and internship opportunities.

Massachusetts Institute of Technology (www.mit.edu)
Graduate Program in Electrical Engineering Computer Science: Artificial Intelligence Research Area
At the Massachusetts Institute of Technology, there are so many laboratories, institutes, and research groups that it can be hard to keep track! Of all the universities represented on this robotics engineering degree ranking, MIT is perhaps the most comprehensive and research-intensive. In fact, the Computer Science and Artificial Intelligence Laboratory (CSAIL) is the largest research center at MIT, and is home to more than 1,000 members, 100 principle investigators, and approximately 50 research groups. Students who are looking to get their foot in the door at this prestigious institution should apply to the graduate program in Electrical Engineering and Computer Science, which – in addition to having close ties with CSAIL – also offers targeted instruction in cognitive robotics, machine learning, and artificial intelligence.

Johns Hopkins University (www.jhu.edu)
Robotics MSE
Johns Hopkins University is best known for its stellar medical programs. What does that have to do with robotics? A lot, apparently! Even a cursory glance at the school's Laboratory for Computational Sensing and Robotics research page and you'll see exactly why. Most of the work going on in JHU robotics – including research in image-guided surgical systems and computer-aided diagnostics – have immediate applications to the medical field. Unsurprisingly, robotics engineering degree students at Johns Hopkins can even choose a "Medical Robotics and Computer Integrated Surgical Systems" specialization track.

Courses in India:
PG Diploma in Machine Learning and AI – Upgrad and IIIT-B: This is an industry-relevant academically-rigorous 11-month on line program covering

Machine Learning and AI concepts in collaboration with IIITB and Uber. It is specifically designed for working professionals with Math/Software Engineering/Statistics/Analytics backgrounds to help them gain practical knowledge and accelerate transition into Advanced Data Science and Machine Learning roles. Course fee is Rs. 2,75,000/-.

Foundations of Artificial Intelligence and Machine Learning, IIIT Hyderabad: Conducted as weekend contact session for a period of 15 weeks, The Machine Learning Lab acts as an umbrella organization at the institute to both strengthen the existing groups and facilitate new activities in related areas. It also acts as a force multiplier in attracting projects and funding from other entities in the government and industry sectors, coordinate research in related domains across different centers of IIIT Hyderabad, as well as in the institute's research collaboration with other academic institutions in the country. Course fee is Rs. 2,75,000/-.

Master of Technology in Artificial Intelligence, University of Hyderabad: M.Tech Artificial Intelligence is also a four-semester course including two semesters of course work and two semesters of project work. This programme is meant for students already well equipped in computing sciences and as such imparts advanced training in all the major areas of artificial intelligence and other emerging technologies, such as machine learning, data mining, etc.

3.16 START-UPS IN INDIA

With a lot of development in AI happening around the globe, India is coming up with equally unique perspectives on the field, as their other developed counterparts. Apparently, in India alone the number of start-ups booming in the space touches 200+, with investments flowing from all major investors.

> **Arya.ai:** A deep learning platform for enterprises to quickly build and scale AI apps and predictive models using Neural Networks, Arya.ai was started in 2013 by Vinay Kumar & Deekshith Marla. With tools like Drag and Drop, it makes it easier to create complex neural networks and its simulation tool expedite the model selection and deployment. The platform has an ecosystem of apps in verticals like Insurance Claims automation, Risk analysis for P&C insurance, Banking Anomaly Detection, Fraud classification; and Manufacturing Energy prediction.

- **Boxx.ai:** Set up in July 2016, Boxx.ai is a software provider that uses Artificial Intelligence to democratize analytics. It uses AI to build products that enable enterprises solve their most critical analytics problems. The products ensure quick results, significant business benefits, all in a pay-per-use pricing. AIDA, the first AI product by Boxx.ai, uses AI to personalize every customer touch-base across channels through real time recommendation, targeted communication and context led app notifications.
- **Cuddle.ai:** It is created as personal AI-driven analyst using sophisticated artificial intelligence and machine learning algorithms. Cuddle is geared to understand the business context, as well as its users' individual responsibility areas. It aims to help business users make more informed decisions through data by delivering time-sensitive and mission-critical insights directly to the mobile devices of business users, in a hyper-personalized manner. Incubated by Fractal Analytics, Cuddle is available on SaaS model and with a monthly per-user license.
- **Embibe:** Founded in 2013 by Aditi Avasthi, it is an AI platform for education outcomes that has traversed its way from being a data oriented company to AI platform for personalization. It offers guaranteed improvement in learning outcomes with behavioural interventions and bite sized learning driven from its unique knowledge tree. Over four years, they2 have fetched data signals space available for insights mining for JEE, NEET, and state boards and foundation exams.
- **Haptik:** Aakrit Vaish launched in Haptik 2014, and is India's first personal-assistant app. Smartphone users, both Apple and Android, can download the Haptik app, which features artificial intelligence. It lets the users plan travel, check-in for flights, book cabs, and set travel reminders. The company believes that with Indians spending 75 percent of their smartphone time on a text or chat interface, they look into improving that experience by leveraging technologies such as artificial intelligence and machine-learning.
- **Netradyne:** It was founded in September 2015 by Avneesh Agrawal and David Julian as a startup pioneering in next gen commercial driver technology, focusing on leveraging artificial intelligence and deep learning to improve road and driver safety. Netradyne's flagship product, Driveri, is a driver assistance and monitoring device whose features include

Quad HD cameras, LTE, GPS, accelerometer, gyrometer, and a Deep Learning processing unit. The device is mounted on the windshield and using its onboard cameras and deep learning processor to analyses the driver performance. If the algorithm recognizes an 'unsafe driving event' it captures the video around that event and uploads it to the Netradyne cloud. This data is then used to create a driver score and the associated video can be used to coach the driver to improve his/her driving habits. Netradyne is using its algorithm and hardware to improve driver safety and reduce accidents by positively modulating driver behavior.

- **Tricog:** Founded by Dr. Charit Bhoraj, Zainul Charbiwala, Udayan Dasgupta and Abhinav Gujjar in Jan 2015 with a belief that 50% of lives lost due to heart disease could be saved if diagnosed quickly. Tricog acquires physiological data and ECG's from clinical and personal health devices and upload them to the cloud. AI powered algorithms process the data and arrive at a diagnosis. A team of specialists work in real-time with the algorithm to verify the diagnosis which is then sent back to assist doctors at remote centers. Tricog has treated over 270,000 patients across 600 remote locations during which time their AI engine has along with the few specialists diagnosed over 14,000 patients with critical heart problems, thus directly saving lives. Tricog is present in over 306 cities/towns across 23 states, in some of the most remote locations in India.

- **Zenatix:** The company is co-founded by Rahul Bhalla, Vishal Bansal, and Amarjeet Singh. It uses advanced machine learning based models to deliver up to 30% energy efficiency to large scale commercial consumers of electricity. The models are built by collecting and analyzing millions of previously unseen data points every day from electrical assets. These models allow for automated and intelligent controls along with predictive and preventive maintenance of these assets, resulting in consistent and comfortable consumer experience in retail and banking sites while delivering high energy savings. With their first product, WattMan, launched in April 2016, it is addressing energy efficiency problem across a range of customers to control the electricity spends and addresses various challenges in an automated manner through its cloud based intelligent controls driven by advanced machine learning algorithms.

- **Sastra Robotics:** It is forerunner in providing cost effective advanced robotics and automation services. It has gained grounds by collaborating with three other pioneering companies in the same market – Switzerland based Cyberbotics, Korea based Dongbu Robot and US based Corobot. to develop and market high end robotic products for wider applications.
- **Systemantics:** It is an Industrial Robotics company which is on a mission to enable widespread adoption of flexible automation in industry. With the experience of having implemented custom turnkey solutions for several clients, the company has accomplished several innovations in robotics and industrial automation such as under water robots and walking machines. Systemantics is focused on manufacture of high-quality industrial robots with design innovations that facilitate affordable solutions and a faster ROI.
- **Gridbots:** Founded in August 2007 by a passionate innovator Mr. Pulkit Gaur (TED Fellow and MIT Young Innovator of the year), the company works in the field of Robotics, Artificial Intelligence and Machine Vision. Gridbots today are a team of more than 30 people – Head quartered in Ahmedabad with sales and support offices in Jaipur, Meerut and Faridabad. Since inception the company has been growing exponentially both in terms of market share and technological capabilities.
- **GreyOrange:** It is a multinational technology firm that designs, manufactures and deploys advanced robotics systems for automation at distribution and fulfillment centers. They combine expertise in robotics, hardware and software engineering to solve operational inefficiencies in warehouse operations.

3.17 ROBOT OLYMPIAD

- **Robot Olympiad World:** WRO brings together young people from all over the world to develop their creativity, design and problem solving skills through challenging and educational robot competitions and activities. The international WRO final takes place once every year in November and teams qualify from national competitions to represent their countries. Costa Rica had hosted WRO2017. Students were asked to come up with innovative solutions using robotic technology to create

Sustainbots i.e. Robots for Sustainablilty. The Challenge for senior group was to make a robot that helps to build a wind farm. The robot would select the best places to build different wind generators of the wind farm in order to ensure maximum efficiency from the generators and no impact on the environment. Team G-Force and team Archimedes of India were winners. Other winners were teams from China, Russia, Hungary, Thailand, Germany and Japan.

> **World Robot Olympiad India:** (WRO India) is one of the largest Robotics competition in India for students between the age group of 9 to 25 years, held since 2006. It is a not-for-profit competition organized by India Stem Foundation. India STEM Foundation is a pioneer in promoting hands-on STEM (Science, Technology, Engineering and Maths) education in India with the intervention of Robotics and research based learning programs for school and college students. India STEM foundation is the affiliate partner of World Robot Olympiad Association in India. World Robot Olympiad™ is an event for science, technology and education which brings together the young people all over the world to develop their creativity and problem-solving skills through challenging and educational robotics competitions. Schools are invited to enter a team of three students to participate in this competition. Participating teams need to create, design and build a robot model that look or behave like a human athlete. The 2017 competition was held at Noida and had categories from preliminary to the final. There were 21 winners in various categories, out of which 11 teams were nominated for international competition.

3.18 CONCLUSION

The question that futurists, AI experts, and people with interest in technology ask as to the time when true AI will be achieved and its consequences. The scientists at the Dartmouth Summer Research Project on Artificial Intelligence in 1956 thought that perhaps two months would be enough to make "significant advances" in a whole range of complex problems of AI. Sixty two years later these problems are not yet solved. Computer vision at the Large Scale Visual Recognition can't answer questions about images very well. In other narrow fields, AI is still catching up to humans. The Loebner Prize is an annual competition in artificial intelligence that awards prizes to the computer programs considered by the judges to be the most human-like. AI has difficulty with these. On the other hand computers are superhuman at chess, and now even at more complex game like Go. AlphaGo's success was much faster than expected. Does this necessarily mean we're closer to general intelligence?

An index of AI has been created and launched as a project of the *One Hundred Year Study on AI at Stanford University* (ai100stanford.edu). It is an open, not-for-profit project to track activity and progress in AI. But this index doesn't attempt to offer a timeline for general intelligence. Michael Woodridge, head of Computer Science at the University of Oxford, says, "The main reason general AI is not captured in the report is that neither I nor anyone else would know how to measure progress." Progress in AI over the next few years is far more likely to resemble a gradual rising tide—as more and more tasks can be turned into algorithms and accomplished by software—rather than the tsunami of a sudden intelligence explosion or general intelligence breakthrough.

Elon Musk and Stephen Hawking have showed concern about artificial intelligence. Now Bill Gates has also leapt into the fray, voicing concern even when chief of research at Microsoft said advanced AI doesn't worry him. This debate of Musk, Hawking, and Gates is about the future of AI when general AI emerges. General artificial intelligence would match and then, very quickly, exceed human intelligence. In his book *Superintelligence*, Nick Bostrom, argues that there are good reasons to believe artificial superintelligence could be very alien and very powerful. Bostrom goes on to say that AI, ironically, may offer the best safeguard. "The idea is to leverage the superintelligence's intelligence, to rely on its estimates of what we would have instructed it to do." So, we should take predictions with a pinch of salt. At the same time, we cannot be unprepared. We don't know how much further there is to go, or how far AI can go. But AI is going somewhere.

REFERENCES
1. Machine Learning by Ethem Alpayoin.
2. Artificial Intelligence by Jerry Kaplan.
3. The Rise of the Robots by Martin Ford.
4. Robotics by Martin Ford.
5. The Industries of the Future by Alec Ross.
6. The Fourth Industrial Revolution by Klaus Schwab.
7. AI: A Modern Approach by Stuart Russel and Peter Norvig.
8. Robotics Wiki. https://en.wikipedia.org/wiki/Robotics
9. Humanoid Robot. https://en.wikipedia.org/wiki/Humanoid_robot
10. Industrial Robot. https://en.wikipedia.org/wiki/Industrial_robo
11. Underwater Robotics. https://en.wikipedia.org/wiki/Underwater_robotics

12. History of Robots https://en.wikipedia.org/wiki/History_of_robots
13. Machine Learning. https://en.wikipedia.org/wiki/Machine_learning
14. Computer Vision. https://www.sciencedaily.com/terms/computer_vision.htm
15. Natural Language Processing in AI. www.expertsystem.com/examples-natural-language-processing-systems-artificial-intelligence

APPENDIX

List of US Universities having Robotics Program:
1. Brown University: http://robotics.cs.brown.edu/
2. Cal Poly Pomona: https://robotics.nasa.gov/students/work.php
3. Caltech: http://www.cms.caltech.edu/research/robotics
4. Carnegie-Mellon University: http://ri.cmu.edu/
5. Colorado School of Mines: http://inside.mines.edu/-MECH-Robotics-Automation-Design
6. Columbia University: http://engineering.columbia.edu/
7. Cornell University: http://robotics.cornell.edu/
8. Dartmouth College: https://engineering.dartmouth.edu/
9. Drexel University: http://drexel.edu/engtech/academics/laboratories/Robotics%20and%20Mechatronics/
10. Florida A&M University/FSU: http://www.eng.famu.fsu.edu/ciscor/academics/
11. Georgia Tech: http://www.robotics.gatech.edu/
12. Indiana University: http://www.indiana.edu/~roboclub/
13. Johns Hopkins University: https://engineering.jhu.edu/fields-of-study/robotics/
14. Kansas State University: http://www.k-state.edu/ksu-robotics/
15. Long Beach City College: http://www.lbcc.edu/Electrical/RoboticsProgram.cfm
16. MIT: https://robotics.mit.edu/
17. New Mexico Institute of Mining & Technology: http://infohost.nmt.edu/~mecheng/ril/david_p.php
18. North Carolina State University: https://www.ece.ncsu.edu/research/crm/
19. Northwestern Polytechnic University: http://www.mccormick.northwestern.edu/robotics/
20. Ohio State University: https://u.osu.edu/robotics/courses/
21. Oregon State University: http://robotics.oregonstate.edu/

22. Portland State University: https://www.pdx.edu/ece/research-laboratories
23. Rice University: http://www.cs.rice.edu/CS/AIRobotics/
24. Southern Illinois University, Edwardsville: http://roboti.cs.siue.edu/
25. Stanford University: http://cs.stanford.edu/group/manips/
26. Tennessee State University: http://www.tnstate.edu/itmrl/
27. Texas A&M University: http://parasol-www.cs.tamu.edu/dsmft/
28. U.S. Air Force Academy: http://www.usafa.af.mil/
29. UC Berkeley: https://robotics.eecs.berkeley.edu/
30. UC San Diego: http://cvrr.ucsd.edu/
31. UC Santa Cruz: https://www.soe.ucsc.edu/departments/computer-engineering/undergraduate/bs-robotics-engineering
32. University of Arizona: http://www.ame.arizona.edu/
33. University of Cincinnati: http://www.robotics.uc.edu/
34. University of Florida: http://www.mil.ufl.edu/
35. University of Hawaii: http://www.eng.hawaii.edu/~asl/
36. University of Houston (Downtown campus): https://www.uhd.edu/academics/sciences/computer-science-engineering-technology/Pages/cset-degree-plan-bs-ciet.aspx
37. University of Houston (Main and Clear Lake campuses): http://www.egr.uh.edu/
38. University of Houston (Main and Clear Lake campuses): http://www.uh.edu/stem/
39. University of Idaho: http://www2.isu.edu/ctech/robotics/
40. University of Maryland: http://www.cs.umd.edu/projects/amrl/
41. University of Maryland: http://www.ssl.umd.edu/
42. University of Massachusetts, Amherst: https://mie.umass.edu/mechatronics-and-robotics-research-laboratory
43. University of Massachusetts, Amherst: https://www.cics.umass.edu/research/area/robotics-computer-vision-and-graphics
44. University of Michigan, Ann Arbor: https://robotics.umich.edu/
45. University of Minnesota: http://www.cs.umn.edu/Research/airvl/
46. University of Missouri: http://engineering.missouri.edu/
47. University of Nebraska-Lincoln: http://engineering.unl.edu/research/robotics-and-mechatronics-lab/

48. University of New Hampshire: http://www.ece.unh.edu/robots/rbt_home.htm
49. University of New Mexico: http://www.mfg.unm.edu/robotics/index.html
50. University of New Mexico: http://pursue.unm.edu/robotics/
51. University of New Mexico: https://losalamos.unm.edu/degrees-certificates/associate-of-applied-science/applied-technologies/robotics.html
52. University of New Mexico: http://engineering.unm.edu/
53. University of Notre Dame: https://cse.nd.edu/
54. University of Notre Dame: https://engineering.nd.edu/
55. University of Oklahoma: http://www.ou.edu/coe.html
56. University of Pennsylvania: http://www.grasp.upenn.edu/
57. University of Rochester: http://www.cs.rochester.edu/users/faculty/brown/lab.html
58. University of South Florida: http://www.csee.usf.edu/robotics/crasar/
59. University of Southern California: http://www-robotics.usc.edu/
60. University of Southern California: https://www.cs.usc.edu/academics/masters/intelligent-robotics
61. University of Tennessee, Knoxville: http://www.cs.utk.edu/~parker/Distributed-Intelligence-Lab/index.html
62. University of Tennessee, Knoxville: http://imaging.utk.edu/
63. University of Texas, Austin: https://robotics.utexas.edu/
64. University of Texas, Dallas: http://www.utdallas.edu/dept/eecs/
65. University of Utah: https://robotics.coe.utah.edu/robotics-track/
66. University of Utah: https://robotics.coe.utah.edu/
67. University of Washington, Bothell: http://faculty.washington.edu/cfolson/mapping.html
68. University of Washington, Seattle: http://brl.ee.washington.edu/
69. University of Washington, Seattle: https://www.cs.washington.edu/research/robotics
70. University of Wisconsin, Madison: http://robotics.engr.wisc.edu/cgi-bin/wikiwp/
71. Utah State University: http://www.csois.usu.edu/
72. Vanderbilt University: https://engineering.vanderbilt.edu/eecs/Research/Robotics.php

73. Vanderbilt University: https://engineering.vanderbilt.edu/eecs/brochure/our-research/robotics.php
74. Vanderbilt University: http://eecs.vanderbilt.edu/CIS/IRL/
75. Villanova University: https://www1.villanova.edu/villanova/engineering/research/Faculty/robotics.html
76. Virginia Tech: http://www.me.vt.edu/research/laboratories/trec/
77. Wellesley University: http://www.wellesley.edu/Physics/robots/studio.html
78. Worcester Polytechnic Institute: https://www.wpi.edu/academics/departments/robotics-engineering

CHAPTER 4
NANOTECHNOLOGY

INTRODUCTION

Generally, builders and makers of structures are focused on building monumental things like pyramids, skyscrapers or aircraft carriers. At the other end of the spectrum, there is another side which is rapidly coming into focus – Nanotechnology. Small is beautiful – the field of nanotechnology aims to build components and even entire machines so small, they approach the atomic level. Tiny machines may cruise our bodies selectively releasing drugs, repairing cells, or hunting pathogens. Nanotechnology may yield materials with amazing new properties. But first the search is on to manipulate matter on the tiniest scales. A new DARPA (Defense Advanced Research Projects Agency) program called Atoms to Product (A2P) hopes to achieve that. The research focuses on developing practical miniaturization and assembly methods at scales 100,000 times smaller than today's most advanced techniques.

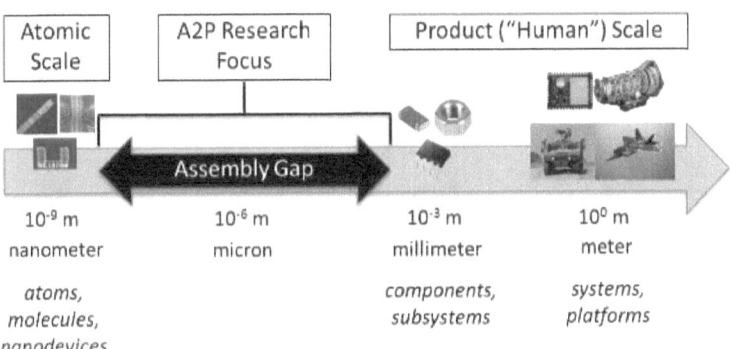

In addition to learning to make better nanoscale components and machines, DARPA is also making materials with useful nanoscale properties on human scales. DARPA programme manager John Main said "If successful, A2P could help enable creation of entirely new classes of materials that exhibit nanoscale properties at all scales, it could lead to the ability to miniaturize materials, processes and devices that can't be miniaturized with current technology, as well as build three-dimensional products and systems at much smaller sizes." Main notes that such assembly naturally occurs in plants and animals, each of which is made up of cells and proteins a million or a billion times smaller than the organism itself. He hopes to enable similar assembly for manmade materials and devices. (www.darpa.mil) A fascinating method of miniaturized fabrication is 3D printing of structures which has become one of the most exciting technologies. While some enthusiasts showcase the technology by producing toys, cars, and even guns in their garage, others look to 3D printing to manufacture the next generation of electronics, whether for mobile applications, medical devices, or wearable computing. Researchers from the Vienna University of Technology have refined a 3D printing technique to make incredibly small structures at record-breaking speeds. The technique, called two-photon lithography, allows the sculpting of intricate objects with micron precision at a rate of 5 meters per second, which is 50,000 times faster than typical processing speeds. The team demonstrated the fabrication from CAD files of St. Stephen's cathedral, the London Tower Bridge, and even a race car that measures about 300 micrometers long, just under the size of a dust mite barely visible to the human eye. Here's the fabricated intricate and detailed cathedral with resolutions at 100 nanometers next to a dust mite

Richard Feynman, the famous inspirational CalTech physicist and bongo-playing eccentric, in his lecture as early as 1959 said about future of miniaturization

"Plenty of Room at the Bottom," and speculated about a world where moving individual atoms would be possible. In other words, if the technology could be developed to make structures with nanoscale precision, the possibilities would be endless. Because of the gap between human scale and the absolute limit that objects can be built at, which is arguably the near-atomic scale, there is plenty of space to build things that perform functions at that size. As lithography techniques become increasingly sophisticated and fast, it's clear that there is a race to the bottom, as technology continues to push to make things smaller. Biotech and nanotech researchers around the world are certainly anxious to get printers in the lab and utilize them for novel experiments. DARPA's work is often of the moonshot variety. When it does work, it tends to go big. Advanced nanotech is in that category. We are closing in on a time when precise, complex engineering at the submicron level is commonplace, and that opens the door to an entirely new world at the bottom. What is astonishing is the potential consequences it could have for society. Suddenly, all one needs is light and raw materials. After all, what is food? It is nothing but combination of carbon, hydrogen, nitrogen, phosphorous, and sulphur. All that is available in abundance. The same elements as listed above could just as easily make insulin, paracetamol, and presumably the superior drugs of the future, too. There's no need to worry about not having medicine as long as we have the recipe and a nanofabricator. Suddenly only three commodities have any value: the raw materials for the nanofabricator many of which, depending on what is to be to made, will be plentiful just from the world around us, the nanofabricators themselves and finally, the blueprints for the things we want to make. In a world where material possessions are abundant for everyone, there will not be a necessity to hoard. In other words, having nanofabricators in place, the problem of non-abundance will be solved. In some ways, it is akin to what a superintelligent AI might make of the human race. In such a world, we too would be different to the one we live today liberated from the drive for survival. Nanotechnology will usher a time in future which will lead to the door of singularity. No human attempts can comprehend what is inside a black hole – a physical singularity. Similarly, inside the veil of this technological singularity, no human attempts at prognostication can really comprehend what the future will look like.

4.1 DEFINITIONS

Nanoscience vs Nanotechnology: The most common definition of nanoscience is "Nanoscience is the study of phenomena and manipulation of materials at atomic, molecular and macromolecular scales, where properties differ significantly from those at a larger scale". Bulk materials (the 'big' pieces of materials we see around us) possess continuous (macroscopic) physical properties. The same applies to micron-sized materials (e.g. a grain of sand). But when particles assume nanoscale dimensions, the principles of classic physics are no longer capable of describing their behaviour (movement, energy, etc.). At these dimensions, the principles of quantum mechanics principles takes over. The same material (e.g. gold) at the nanoscale can have properties (optical, mechanical and electrical) which are very different from (and even opposite to) the properties the material has at the macroscale (bulk). Nanoscience is an interdisciplinary science, which means that it involves concepts of more than one discipline, such as chemistry, physics, etc. and further expands the borders by adding biology and biochemistry to the mix. Nanoscience is thus a 'horizontal-integrating interdisciplinary science that cuts across all vertical sciences and engineering disciplines.

The application of nanoscience to practical devices is called Nanotechnologies which are based on the manipulation, control and integration of atoms and molecules to form materials, structures, components, devices and systems at the nanoscale. Nanotechnologies are the application of nanoscience especially to industrial and commercial objectives. All industrial sectors rely on materials and devices made of atoms and molecules thus, in principle, all materials can be improved with nanomaterials, and all industries can benefit from nanotechnologies.

Nanometre Scale: The term nanotechnology comes from the combination of two words viz the Greek numerical prefix nano referring to a billionth and the word technology. As an outcome, Nanotechnology or Nanoscaled Technology is generally considered to be at a size below 0:1 m or 100 nm (a nanometer is one billionth of a meter, 10^9 m). Nanoscale science (or nanoscience) studies the phenomena, properties, and responses of materials at atomic, molecular, and macromolecular scales, and in general at sizes between 1 and 100 nm. In this scale, and especially below 5 nm, the properties of matter differ significantly (i.e., quantum-scale effects play an important role) from that at a larger particulate scale. Nanotechnology is then the design, the manipulation, the building, the production and application, by controlling the shape and size, the properties, responses and functionality of structures, and devices and systems of the order of less than 100 nm. Nanotechnology per se was first discussed at American Physical Society meeting at the California Institute of Technology on December 29, 1959 by renowned physicist Richard Feynman in his talk "There's Plenty of Room at the Bottom". Professor Norio Taniguchi, over a decade later coined the term nanotechnology. National Nanotechnology Initiative (NNI) defines nanotechnology as the manipulation of matter having at least one dimension varying between 1 to 100 nanometers, where unique quantum mechanical effects occur at this quantum-realm scale. Matter such as gases, liquids, and solids can exhibit unusual physical, chemical, and biological properties at the nanoscale. Some nanostructured materials are stronger or have different magnetic properties compared to other forms or. Others are better at conducting heat or electricity and may become more chemically reactive or reflect light better or change color as their size or structure is altered. Thus, nanotechnology may be able to create many new materials and devices with a vast range of applications, such as in medicine, electronics, biomaterials energy production, and consumer products. On the other hand of the spectrum, nanotechnology may raise concerns about the toxicity and environmental impact of nanomaterials. How small is "nano?"

- There are 25,400,000 nanometers in one inch
- A human hair is approximately 80,000–100,000 nanometers wide
- A single gold atom is about a third of a nanometer in diameter
- On a comparative scale, if the diameter of a marble was one nanometer, then diameter of the Earth would be about one meter
- A DNA molecule is 1–2 nm wide

4.2 WHAT IS SPECIAL ABOUT NANO

There are various reasons why nanoscience and nanotechnologies are so promising in materials, engineering and related sciences. First, at the nanometre scale, the properties of matter, such as energy, change. This is a direct consequence of the small size of nanomaterials, physically explained as quantum effects. The consequence is that a material (e.g. a metal) when in a nano-sized form can assume properties which are very different from those when the same material is in a bulk form. For instance, bulk silver is non-toxic, whereas silver nanoparticles are capable of killing viruses upon contact. Properties like electrical conductivity, colour, strength and weight change when the nanoscale level is reached. The same metal can become a semiconductor or an insulator at the nanoscale level. The second exceptional property of nanomaterials is that they can be fabricated atom by atom by a process called bottom- up. The information for this fabrication process is embedded in the material building blocks so that these can self-assemble in the final product. Finally, nanomaterials have an increased surface-to-volume ratio compared to bulk materials. This has important consequences for all those processes that occur at the surface of a material, such as catalysis and detection.

The use of nanostructured materials dates back to fourth century AD when Romans were using nanosized metals to decorate glasses and cups. One of the first known, and most famous example, is the Lycurgus cup that was fabricated from nanoparticles (NPs) from gold and silver that were embedded in the glass. The cup depicts King Lycurgus being dragged to the underworld.

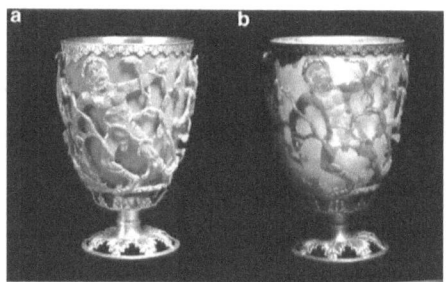

Under normal lighting, the cup appears green. However, when illuminated from within, it becomes vibrant red in colour. In that cup, as well as in the famous stain glass windows from the tenth, eleventh, and twelfth centuries, metal NPs account for the visual appearance. This ability of gold as well as of other noble metals and semiconductors relies on quantum confinement. According to this theory electrons are confined in all three dimensions causing matter to behave completely different in terms of its optical and electronic properties. When the dimension of a material approaches the electron wavelength in one or more dimensions, quantum mechanical characteristics of the electrons that are not manifest in the bulk material can start to contribute to or even dominate the physical properties of the material. Besides quantum size effects, the nanomaterials behaviour is different due to surface effects which dominate as nanocrystal size decreases. For metals, the electron mean free path (MFP) determines the thermal and electrical conductivity and affects the colour of the metal. For most of the metals, MFP is of the order of 5–50 nm. Reducing further this threshold, the electrons begin to scatter off the crystal surface, and the resistivity of the particles increases. For very small metal particles, the conduction and valence bands begin to break down into discrete levels. For gold particles, this causes a change in colour from red to orange at sizes around 1.5 nm.

4.3 NANOSCIENCE AND NANOTECHNOLOGY IN NATURE

We see hundreds of examples of nanoscience under our eyes daily, from geckos that walk upside down on a ceiling, apparently against gravity, to butterflies with iridescent colours, to fireflies that glow at night. Nature provide some outstanding solutions to complex problems in the form of fine nanostructures with which precise functions are associated. A short list of some natural nanomaterials is given below.

- **Nanoparticles from natural erosion and volcanic activity** are part of our mineral world since they are naturally produced during erosion and volcanic explosions.
- **Minerals**, such as clays, are nanostructured and are a type of layered silicate characterised by a fine 2D crystal structure. Among clays, mica has been the most studied and is made up of large sheets of silicate held together by relatively strong bonds. The fine nanostructure of clays determines their properties. When water is added, the clay swells. The volume change is several times the original volume due to the 'opening' up of the layered structure by the water molecules that replace the cations. Clay swelling is a significant factor in soil stability and taken into account when building roads etc.
- **Natural colloids**, such as milk, blood, fog, gelatin, nanoparticles are dispersed in the medium but do not form a solution, rather a colloid. All these materials have the characteristic of scattering light and often their colour (as in the case of blood and milk) is due to the scattering of light by the nanoparticles that make them up.

- **Mineralised natural materials**, such as shells, corals and bones are formed by calcium carbonate crystals that self-assemble together with other natural materials, such as polymers, to form fascinating three-dimensional architectures. For instance, a shell is grown by a layer of cells that first lays down a coating of protein supported by a polysaccharide polymer like chitin. The proteins act like a nano-assembly mechanism to control the growth of carbon carbonate crystals. Around each crystal remains a honeycomb-like matrix of protein and chitin. This relatively 'flexible envelope' is fundamental for the mechanical properties of the shell and mitigates cracking. The size of each crystal is around 100 nm. The result is that the nacre of mollusc shells has extraordinary physical properties (strength, resistance to compression, etc.).
- **Materials like skin, claws, beaks, feathers, horns, hair** are made largely of very flexible proteins like keratin, elastin and collagen. Keratins have a large glycine and alanine content. This leads to β-sheets that can bond strongly one with another in an aligned fashion. Fibrous keratin molecules can twist around each other to form helical intermediate laments. Similarly, collagen (not related to keratin in terms of primary structure) has a high percentage of glycine and forms flexible triple-helix structures. In addition to intra and inter- molecular bonds, keratins have numerous cysteins that can form stable disulphide bonds. The amount of cysteins in the protein determines the strength and rigidity of the material. Materials like nails, hooves and claws have a higher percentage of cysteins.
- **Paper and cotton** are both made mainly of cellulose. The high strength, durability and absorbency of cotton are due to the nanoscale arrangement of the fibres.
- **Insect wings and opals:** The colours seen in opals and butterflies are directly related to their fine structure, which reveals packed nanostructures that act like a diffraction grid and induce iridescence. In the case of opals, this is due to packed silica spheres in the nanometre range, uniform in size and arranged in layers. Butterflies often owe the colour of their wings to pigments that absorb specific colours; in some species, such

as the beautiful Morphorhetenor, colours are due to the presence in the wings of nanostructures which are photonic crystals. The wings also exhibit iridescence: the shift in colour of an object when observed at different angles. The effect can easily be seen by observing a music CD. Iridescence is a 'physical colour' and it results from the interaction of light with the physical structure of the surface. To interact with visible light, those structures must be nano-sized (the visible spectrum corresponds to wavelengths between 380 and 750 nm). The interaction of light with this nano-rough surface can lead to constructive or destructive interference. The colour, intensity and angles of iridescence depend on the thickness and refractive index of the substrate, and on the incident angle and frequency of the incident light. In materials like opals, natural iridescence is observed, due to packed silica spheres in the nanometre range, uniform in size and arranged in layers. This provides appropriate conditions for interference. In the case of butterflies and moths, the iridescence is produced in a peculiar way. Scientists have studied the structure of the wings of Morpho-rhetenor, (a neotropical butterfly found in French Guiana, Brasil, Peru) and have found that these are formed by rows of scales arranged like tiles on a roof. Each scale is about 70 x 200 μm and has a smaller structure on its surface, a very intricate and highly ordered nanometre organisation of ridges. Each ridge is about 800 μm wide. The spaces between them form a natural photonic crystal that can generate constructive and destructive interference. The SEM analysis of the cross-section of the ridges on the wings shows and an even more intricate structure that looks like fir trees

- **Spider silk** is the material with the greatest known strength, about five times that of steel of the same weight. The extraordinary properties of spider silk are due to the proteins that make up the silk and its supramolecular organisation which is at the nanoscale level.
- **Lotus leaves**, Nelumbonucifera, is a native Asian plant which has the distinct property of being able to maintain its leaves particularly clean even if its natural habitat is muddy. For this reason, this plant is considered sacred in some cultures and a sign of purity. The leaves of the lotus plant have the outstanding characteristic of totally repelling

water because they are superhydrophobic. The consequence is that water droplets roll of the leaf surface and, in doing so, drag dirt away from it. This effect, 'self-cleaning', renders the lotus leaf clean and resistant to dirt. The same effect is found in other leaves such as those of nasturtium (Tropaeolum) and some Cannas, and in some animals such as the water strider.

> **Geckos' feet** structure is an amazing example of the relationship between function and nanostructure. The ability of geckos to walk upside down, against gravity, even on wet or dirty surfaces, is intimately connected to the nanostructure of their feet. As it walks, a gecko does not secrete any sticky substance, and its feet do not have any suction-like features (even at microscopic sizes). The reason for the gecko's amazing properties lies in the nanostructures that are present on its feet. The gecko foot has a series of small ridges called scansors which contain numerous projections called setae. Each seta is about 100 μm long and has a diameter of about 5 μm. There are about half a million of these setae on the foot of a gecko. Each seta is further subdivided into about a thousand 200 nm-wide projections called spatulae. As a result, the total surface area of the gecko's feet is enormous. The gecko spatulae are very exible, so they essentially mould themselves into the molecular structure of any surface. The result is a strong adhesion which is entirely due to van der Waals forces. A single seta can resist a force of 200 μN, or approximately 10 atmospheres of stress. The gecko case is thus a very good example of the effect of large surface area on small forces. Another very interesting property of geckos is that their feet don't get dirty as they walk, even if they walk on a surface covered with sand, dirt, water, etc. Their feet stay clean even on dirty surfaces and full adhesion is maintained. The phenomenon has been investigated and it was found that the feet remain clean because it is more energetically favourable for particles to be deposited on the surface than to remain adhering to the gecko spatulae. If a gecko walks over a dirty surface, it takes only few steps for its feet to be totally clean again, and adhesion is not compromised.

> **Learning from nature** are of interest not only to understand (and appreciate) the amazing properties of biological nanomaterials but also to gather inspiration for the design and engineering of new materials with advanced properties. The physical origins of the remarkable properties of many biological materials are due to complex, often hierarchical structures. They are characterised by a surprising level of adaptability and multi-functionality. These materials can provide a model for designing radically improved artificial materials for many applications, such as solar cells, fuel cells, textiles, drug delivery systems, etc. What is even more inspiring is the notion that in nature, some very simple laws such as:
>   > Nature runs on sunlight and uses only the energy it needs. Natural nanomaterials are extremely energy efficient!
>   > Nature fits form to function and recycles everything, waste products are minimised in nature!
>   > Nature rewards cooperation although it encourages diversity and local expertise.

The field of materials engineering devoted to trying to fabricate artificial materials that mimic natural ones is conventionally called biomimetics. Nanoscience is a fundamental component of biomimetics.

4.4 NANOMANUFACTURING

Methods for fabricating nanomaterials can be generally subdivided into two groups: top-down methods and bottom-up methods. In the first case, nanomaterials are derived from a bulk substrate and obtained by the progressive removal of material, until the desired nanomaterial is obtained. A simple way to illustrate a top-down method is to think of carving a statue out of a large block of marble. Printing methods also belong to this category. Bottom-up methods work in the opposite direction: the nanomaterial, such as a nanocoating, is obtained starting from the atomic or molecular precursors and gradually assembling it until the desired structure is formed. The method resembles building with Lego bricks. Two requisites are fundamental in both the cases. They are control of the fabrication conditions such as energy of the electron beam and control of the environment conditions like presence of dust, contaminants, etc. For these reasons, nanotechnologies use highly sophisticated fabrication tools that are mostly operated in a vacuum in clean-room laboratories.

Top-down Method: These fabrication methods used in nanotechnologies are derived from the fabrication methods used in the semiconductor industry to fabricate the various integrated circuits. Collectively called lithography, they use a light or electron beam to selectively remove micron-scale structures from a precursor material called resist. Nowadays, it is possible to obtain single features below 100 nm (the transistors in latest generation processors are about 45 nm). Top-down fabrication reduces large pieces of materials all the way down to the nanoscale, like someone carving a model airplane out of a block of wood. This approach requires larger amounts of materials and can lead to waste if excess material is discarded. Michelangelo was a top-down artist. He took one big, raw block of Carrara marble and after years of chiselling away produced a spectacular statue like David. In the process he reduced the original block of marble to half its original volume and left the other half as waste. This is the nanotechnology equivalent of lithography and other top-down methods where a block of material is taken and bits and pieces are removed until desired shape and size is achieved. In the process relatively much energy is used, often quite a bit of waste is generated, need a lot of patience as this process is relatively slow and often the results are quite unique and not easily replicable.

Therefore, in the semiconductor industry, nanostructures are routinely fabricated. Some common methods are given below.

Conventional Lithography: Lithography includes various fabrication techniques that use the principle of transferring an image from a mask to a receiving substrate. Three successive steps of a typical lithographic process are (i) coating a substrate (Si wafer or glass) with a sensitive polymer layer (called resist); (ii) exposing the resist (it is proprietary mixture of a polymer and other small molecules that have been specially formulated for a given for a given lithography technology) to light, electrons or ion beams; (iii) developing the resist image with a suitable chemical (developer), which reveals a positive or negative image on the substrate depending on the type of resist used. In conventional microfabrication used in the semiconductor industry, the next step after lithography is pattern transfer from the resist to the underlying substrate. This is achieved through a number of transfer techniques, such as chemical etching and dry plasma etching.

Photolithography: Uses light (UV, deep-UV, extreme-UV or X-ray) to expose a layer of radiation-sensitive polymer (photoresist) through a mask. The mask is a quartz or glass plate containing information about the feature to be printed.

The resolution of contact mode lithography is typically 05–08 μm when UV light (360–460 nm) is used. Higher resolutions cannot be achieved due to the inability to reduce the gap between the mask and the flat substrate below approximately 1 µm, even when elaborate vacuum systems are used to hold the two parts together. To produce patterns with higher resolution, projection photolithography or 'next-generation photolithography' techniques (i.e. extreme UV and X-ray photolithography) need to be employed. These technologies use very expensive equipment and, therefore, their use is limited to selected applications.

Scanning Lithography: Energetic particles such as electrons and ions can be used to pattern appropriate resist (photosensitive material) lms leading to features with nanometre resolution. When using electrons, the technology is called electron beam lithography (e-beam), whereas when ions are used the technology is called focused ion beam lithography. Finally, a recently established technology uses nanometre scanning probes for patterning resist lms and is therefore referred to as Scanning Probe Lithography (SPL). This technology has been extended to the deposition of a nanoquantity of material.

E-beam Lithography: In a typical e-beam lithography process, a tightly focused beam of electrons scans across the surface of an electron-sensitive resist. The main advantage of e-beam lithography over photolithography is its high resolution: patterns with features as small as 50 nm can be routinely generated.

Soft Lithography: Soft lithography is the name for a number of techniques that fabricate and use a soft mould prepared by casting a liquid polymer precursor against a rigid master. These methods have been developed specifically for making large-scale micro and nanostructures with equipment that is easier to operate compared to those used in conventional lithography cheaper and also available in biological laboratories.

Nanocontact Printing: Microcontact printing is useful for patterning features with lateral dimension of 500 nm or larger. One of the major challenges for µCP (microcontact printing) has been to achieve the capability to print with high resolution with lateral dimension lower than 100 nm. This has recently been achieved by improving the stability of the PDMS (polydimethylsiloxane) and it is now possible to print features as small as 50 nm. This printing method, which uses harder stamps, is called nanocontact printing (nCP).

Nano-imprint Lithography: The concept of nano-imprint lithography is to use a hard master with a 3D nanostructure to mould another material, which assumes its reverse 3D structure. The method is the equivalent of embossing at the nanoscale and requires specialised equipment.

Nanosphere Lithography: In nanosphere lithography, an ensemble of nanospheres ordered on a surface is used as a mask. The nanospheres are dispersed in a liquid and a droplet placed on a surface and left to dry. Depending on the surface properties (e.g. charge) and media used in the colloid, the nanosphere will self-assemble in an ordered pattern.

Colloidal Lithography: Colloidal lithography shares the same principle of nanosphere lithography by using a colloid as a mask for the fabrication of nanostructures on surfaces. In this method, electrostatic forces are employed to obtain short-range ordered arrays of nanospheres on the surface. The array can then be used to create a number of different nanostructures.

Scanning Probe Lithography: Scanning probe microscopy (i.e. scanning tunnelling microscopy STM, atomic force microscopy AFM, etc.) uses small (<50 nm) tips to image surfaces with atomic resolution. This ability suggests opportunities for their use in generating nanostructures and nanodevices. In this form, they are referred to as Scanning Probe Lithography (SPL), which uses the tip of an AFM to selectively remove certain areas on a surface and Dip Pen Nanolithography, which, similarly, uses the AFM tip to deposit material on a surface with nanometre resolution.

Writing 'atom by atom': A particular feature of an STM is that it can be used for more than just to visualise atoms. Twenty years ago, researchers at IBM were able to demonstrate that they could use the STM tip to carefully move atoms on a surface and write the company logo with atoms.

If one were to write using atoms, letters would be around 1 nm each. With letters of this size, the whole of Encyclopaedia Britannica could be reproduced in an area as small as the tip of a human hair (10^{-4} m^2). Indeed, with letters of this size all the world's books would fit on a single A4 sheet, but it would take incredibly long to write, and in addition they could only be read with the STM. Nevertheless, the ability to move individual atoms using an STM has great potential for the future generation of data storage devices. The STM allows a material to be built atom by atom independently of its chemistry and physics. This can lead to new materials that most likely have completely new properties.

The process is still very slow, since the atoms can be moved only manually, and this must be one atom at a time.

Bottom-up Method

Bottom-up methods can be divided into gas-phase and liquid-phase methods. In both cases, the nanomaterial is fabricated through a controlled fabrication route that starts from the single atom or molecules:

- Gas-phase methods: these include plasma arcing and chemical vapour deposition;
- Liquid phase: the most established method is sol-gel synthesis; molecular self-assembly is emerging as a new method.

Plasma Arcing: This is the most common method for fabricating nanotubes. The method uses a plasma which is an ionised gas. A potential difference is placed between two electrodes and the gas in between ionises. These positively changed ions pass to the other electrode, pick up electrons and are deposited to form nanotubes. Plasma arcing can also be used to deposit nanolayers on surfaces rather than making new structures. Plasma arcing is complementary to chemical vapour deposition.

Chemical vapour Deposition: In this method, the material to be deposited is first heated to its gas form and then allowed to deposit as a solid on a surface and operation is done in vacuum. This process is routinely used to make nanopowders of oxides and carbides of metals if carbon or oxygen are present with the metal. The method can also be used to generate nanopowders of pure metals though not so easy to do. Chemical vapour deposition is often used to deposit a material on a flat surface. When a surface is exposed to a chemical vapour, the first layer of atoms or molecules that deposit on the surface can act as a template on which material can grow. The structures of these materials are often aligned. During deposition, a site for crystallisation may form in the depositional axis. As a result, aligned structures start to grow vertically. This is therefore an example of self-assembly.

Molecular beam Epitaxy: This is a very sophisticated evaporation method in which molecular beams interact on a heated crystalline substrate under ultra-high vacuum conditions to produce a single crystal lm. Molecular Beam Epitaxy makes it possible to fabricate crystals one atomic layer at a time. The growth process is highly controlled to avoid contaminants being introduced during the crystal growth. MBE is used for the fabrication of numerous important devices such as light-emitting diodes, laser diodes, field effect transistors, read-write heads for computer drives and more.

Sol-gel Synthesis: This method is carried out in the liquid phase. It is a useful self-assembly process for fabricating nano-particles as well as nanostructured surfaces and three-dimensional nanostructured materials such as aerogels. A 'sol' is a type of colloid in which a dispersed solid phase is mixed into a homogeneous liquid medium. An example of a naturally occurring sol is blood. The sol-gel process involves the evolution of networks through the formation of a colloidal suspension and gelation of the sol to form a network in a continuous liquid phase. The sol-gel process involves four steps. First, the hydrolysis reaction, in which the -OR group is replaced with an -OH group. The hydrolysis reaction can occur without a catalyst but is more rapid and complete when catalysts are used. As in any hydrolysis reaction, the catalyst can be a base (NaOH or NH3) or an acid (HF or CH3COOH). After hydrolysis, the sol starts to condense and polymerise. This leads to a growth of particles which, depending on various conditions such as pH, reach dimensions of a few nanometres. The condensation/ polymerisation reaction is quite complex and involves many intermediate products, including cyclic structures. The particles then agglomerate: a network starts to form throughout the liquid medium, resulting in thickening, which forms a gel. The sol-gel process is very commonly used to make silica gels. Other type of gels can also be formed: aluminosilicate gels are special because they form tubular structures. These types of nanostructures are known to be good adsorbents of anions such as chloride, chlorate, sulphate and phosphate.

Molecular Self-assembly: Self-assembly is the 'fabrication tool' of nature: all natural materials, organic and inorganic, are produced through a self-assembly route. In natural biological processes, molecules self-assemble to create complex structures with nanoscale precision. Examples are the formation of the DNA double helix or the formation of the membrane cell from phospholipids. In self-assembly, sub-units spontaneously organise and aggregate into stable, well-defined structures through non-covalent interaction. This process is guided by information that is coded into the characteristics of the sub-units and the final structure is reached by equilibrating to the form of the lowest free energy.

It is now possible for extremely small objects but using a computer-assisted program to actually build a device, such as an electronic circuit, atom by atom, through a self-assembly program is still in the domain of future. However it is possible to fabricate some very small components, such as nanowire, of an integrated circuit through a self-assembly process.

DNA Nanotechnology: DNA nanotechnology exploits the structural motifs and self-recognition properties of DNA to self-assemble pre-designed nanostructures in a bottom-up approach. Two and three-dimensional structures have been fabricated using this self-assembly method. Recently, the revolutionary DNA origami method was developed to build two-dimensional addressable DNA structures of arbitrary shape that can be used as a platform to arrange nanomaterials with high precision and specificity. Researchers at the Centre for DNA Nanotechnology at Aarhus University have developed a software package to facilitate the design of DNA origami structures and it was initially applied in the design of the dolphins in the former logo of Aarhus University. DNA nanotechnology represents one of the latest developments in nanotechnology. It has applications for the fabrication of nanoguides (e.g. waveguides), sensors (for diagnostic and imaging), logic gates, drug release, nano-motors and electronics (wires, transistors). It could lead to bottom-up electronics and DNA computing, which could become the computing of the future.

4.5 TYPE OF NANOMATERIALS

As already seen, nanotechnology generally refers to designing and making anything whose use depends on specific structure at the nanoscale – generally taken as being 100 nanometres or less. It includes devices or systems made by manipulating individual atoms or molecules, as well as materials which contain very small structures. Nanomaterials are usually considered to be materials with at least one external dimension that measures 100 nanometres or less or with internal structures measuring 100 nm or less. They may be in the form of particles, tubes, rods or fibers. The nanomaterials that have the same composition as known materials in bulk form may have different physico-chemical properties than the same materials in bulk form, and may behave differently if they enter the body. They may thus pose different potential hazards. The number of products produced by nanotechnology or containing nanomaterials entering the market is increasing. Current applications include healthcare, electronics, cosmetics, textiles, information technology and environmental protection. For example, nanosilver is appearing in a range of products, including washing machines, socks, food packaging, wound dressings and food supplements. Nanomaterials can broadly be categorized as follows:

Engineered nanomaterials are deliberately engineered and manufactured by humans to have certain required properties.

Incidental Nanomaterials are incidentally produced as a byproduct of mechanical or industrial processes. Sources of incidental nanoparticles include vehicle engine exhausts, welding fumes, combustion processes from domestic solid fuel heating and cooking. Incidental atmospheric nanoparticles are often referred to as ultrafine particles, and are a contributor to air pollution.

Natural nanomaterials are natural biological systems. The structure of foraminifera (mainly chalk) and viruses, the wax crystals covering a lotus or nasturtium leaf, spider and spider-mite silk,the blue hue of tarantulas, the "spatulae" on the bottom of gecko feet, some butterfly wing scales, natural colloids (milk, blood), horny materials, paper, cotton, nacre, corals, and even our own bone matrix are all natural organicnanomaterials. Natural inorganic nanomaterials occur through crystal growth in the diverse chemical conditions of the Earth's crust. For example, clays display complex nanostructures due to anisotropy of their underlying crystal structure, and volcanic activity can give rise to opals, which are an instance of a naturally occurring photonic crystals due to their nanoscale structure. Natural nanoparticles also include combustion products forest fires, volcanic ash, ocean spray, radioactive decay of radon gas and can also be formed through weathering processes of metal or anion containing rocks, as well as at acid mine drainage sites. Nanomaterials are generally categorized in different types. A *nanoparticle* is defined a nano-object with all three external dimensions in the nanoscale, whose longest and the shortest axes do not differ significantly. A *nanofiber* has two external dimensions in the nanoscale, with nanotubes being hollow nanofibers and *nanorods* being solid nanofibers. A *nanoplate* has one external dimension in the nanoscale, and if the two larger dimensions are significantly different it is called a nanoribbon. For nanofibers and nanoplates, the other dimensions may or may not be in the nanoscale but must be significantly larger.

Images of few nanomaterials are given below:

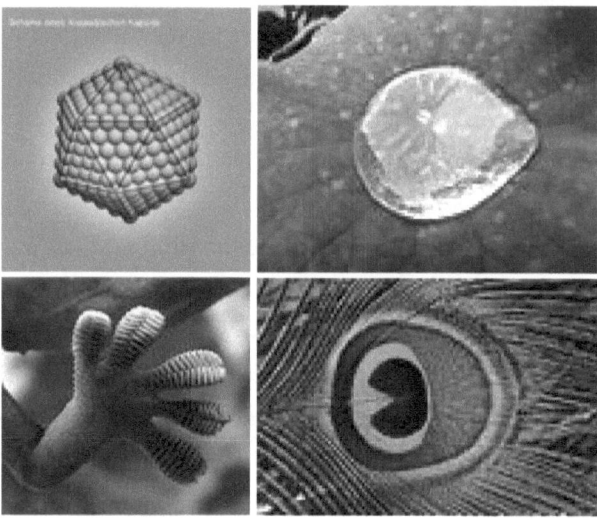

Fullerenes: Fullerenes are a class of allotropes of carbon which conceptually are graphene sheets rolled into tubes or spheres. These include the carbon nanotubes or silicon nanotubes which are of interest both because of their mechanical strength and electrical properties. They are spherical carbon-cage molecules with sixty or more carbon atom having measurement of about 0.7–1.5 nm in diameter. Fullerenes are studied because of their unusual properties. For potential medical use, they are strong antioxidants and could also bind specific antibiotics to the structure to target resistant bacteria and even target certain cancer cells such as melanoma. Heat resistance and superconductivity are some of the more heavily studied properties of fullerenes in mechanical engineering. A common method used to produce fullerenes is to send a large current between two nearby graphite electrodes in an inert atmosphere. The resulting carbon plasma arc between the electrodes cools into sooty residue from which many fullerenes can be isolated.

Carbon Nanotubes: Carbon nanotubes (CNTs) are quite different from fullerene-type materials having rather different properties. CNTs are produced in various ranges from single to tens of nanometers in diameter and several micrometers in length. They have outstanding mechanical and electronic properties and are good thermal conductors. The tensile strength, or breaking strain of CNTs is 6–7 times that of steel. They are among the stiffest and strongest fibers known. Some CNTs are the most efficient electrical conductors ever made, while others behave more like silicon. These properties, coupled with the lightness of carbon nanotubes, give them great potential for use in reinforced composites, nanoelectronics, sensors and nanomechanical devices. Other uses for nanotubes:

- Chemical and Genetic Probes. A nanotube-tipped atomic force microscope can trace a strand of DNA and identify chemical markers that reveal which of several possible variants of a gene is present in the strand.
- Mechanical memory (non-volatile RAM). A screen of nanotubes laid on support blocks has been tested as a binary memory device, with voltages forcing some tubes to contact (the "on" state) and others to separate ("off").
- Field Emission Based Devices. Carbon Nanotubes have been demonstrated to be efficient field emitters and are currently being incorporated in several applications including flat-panel display for television sets or computers or any devices requiring an electron producing cathode such as X-ray sources.
- Nanotweezers. Two nanotubes, attached to electrodes on a glass rod, can be opened and closed by changing voltage. Such tweezers have been used to pick up and move objects that are 500 nm in size.
- Supersensitive Sensors. Semiconducting nanotubes change their electrical resistance dramatically when exposed to alkalis, halogens and other gases at room temperature, raising hopes for better chemical sensors. The sensitivity of these devices is 1,000 times that of standard solid state devices.
- Hydrogen and Ion Storage. Nanotubes might store hydrogen in their hollow centers and release it gradually in efficient and inexpensive fuel cells. They can also hold lithium ions, which could lead to longer-lived batteries. make fuel cells economical. The work with lithium ions is still preliminary.
- Sharper Scanning Microscope. Attached to the tip of a scanning probe microscope, nanotubes can boost the instruments' lateral resolution by a factor of 10 or more, allowing clearer views of proteins and other large molecules.

➢ Superstrong Materials. Embedded into a composite, nanotubes have enormous resilience and tensile strength and could be used to make materials with better safety features, such as cars with panels that absorb significantly more of the force of a collision than traditional materials, or girders that bend rather than rupture in an earthquake.

Graphene: Discovered in 2004, graphene is a flat one-atom thick sheet of carbon. It was long considered that a free-standing form of planar grapheme was impossible as sheet always would roll up. Discovery of this two dimensional crystals has made scientists excited as the material is only one atom thick, the third dimension, height, is considered to be zero and opens up a whole new class of materials with novel electronic, optical and mechanical properties.

Nanoparticles: Metal oxide ceramic, metal, and silicate nanoparticles constitute the most common of the new generation of nanoparticles. Beside them, a substance called chitosan for example, used in hair conditioners and skin creams, has been made in nanoparticle form to improve absorption.

Quantum Dots: Just as carbon nanotubes are often described as the new plastics, so quantum dots are defined as the ball bearings of the nano-age. Quantum dots are like "artificial atoms". They are 1 nm structures made of materials such as silicon, capable of confining a single electron, or a few thousand, whose energy states can be controlled by applying a given voltage. In theory, this could be used to fulfil the alchemist's dream of changing the chemical nature of a material, making lead emulate gold, for example. One more likely set of possible applications exploits the fact that quantum dots can be made to emit light at different wavelengths, with the smaller the dot the bluer the light. The dots emit over a narrow spectrum making them well suited to imaging, particularly for biological samples. Currently, biological molecules are imaged using naturally fluorescent molecules, such as organic dyes, with a different dye attached to each kind of molecule in a sample. But the dyes emit light over a broad range of wavelengths, which means that their spectra overlap and only about three different dyes emit light over a broad range of wavelengths, which means that their spectra overlap and only about three different dyes can be used at the same time. With quantum dots, full-colour imaging is possible because large numbers of dots of different sizes can be excited by a light source with a single wavelength. The wide range of colors that can be produced by quantum dots also means they have great potential in security. They could, for example, be hidden in bank notes or credit cards, producing a unique visible image when exposed to ultraviolet light. Quantum dots are also possible materials for making ultrafast, all-optical switches and logic gates that work faster than 15 terabits a second. For comparison, the Ethernet generally can handle only 10 megabits per second. Biologists are experimenting with composites of living cells and quantum dots. These could possibly be used to repair damaged neural pathways or to deliver drugs by activating the dots with light.

4.6 IMAGING OF NANOMATERIALS

'Seeing is believing'. Therefore, imaging of nanomaterials is an essential part of nanoscience and nanotechnologies. Imaging in nanoscience does not just mean to create an image. Scientists nowadays have access to a variety of truly amazing instruments that allow them to see objects at the nanoscale. This dream came true in the mid 1980s when a revolutionary instrument, the scanning tunnelling microscope, was invented and shortly after, the atomic force microscope. As a matter of fact, it was the invention of these instruments that truly opened the doors to the nanoworld. Once scientists were able to see nanoscale objects, they started to be able to analyse them, understand their behaviour, and imagine ways of manipulating them. In general, two fundamental types of characterisation methods exist: imaging by microscopy and analysis by spectroscopy. The methods employed have been developed specially to meet the characterisation needs of nanomaterials.

Microscopy Methods: An optical microscope, the oldest and simplest microscope, uses visible light and a system of lenses to magnify images of small samples. The resolution limit of an optical microscope is governed by the wavelength of visible light with wavelengths between 400 and 700 nm. The resolving power of an optical microscope is around 0.2 μm or 200 nm. Single objects smaller than this limit are not distinguishable: they are seen as 'fuzzy objects'. This is known as the 'diffraction limit' of visible light. In order to overcome the limitations set by the diffraction limit of visible light, other microscopes have been designed which use electron beams to illuminate the sample. Electron microscopes have much greater resolving power than light microscopes that use electromagnetic radiation and can obtain much higher magnifications of up to two million times. Both electron and light microscopes have resolution limitations, imposed by the wavelength of the radiation used.

There are various types of electron microscopes, such as the scanning electron microscope (SEM) and or the transmission electron microscope (TEM). Conceptually, these microscopes are similar to an optical microscope in the sense that they use radiation to visualise a sample: photons in the case of an optical microscope, and electrons in the case of electron microscopes. In 1981, a totally new concept of imaging was introduced by Binning and his co-workers from IBM. They used a small metal tip placed at a minute distance from a conducting surface. When the two are placed very close together, without actually touching, a bias between the two can allow electrons to tunnel through the vacuum between them. This creates a tunnelling current, which can be measured and which is a function of the electron density on the surface. Electron density is the probability of finding an electron in a particular place: there is high electron density around the atoms and bonds in molecules. This type of microscope is called the Scanning Tunnelling Microscope (STM). Variations in current as the probe passes over the surface are translated into an image. The STM can create detailed 3D images of a sample with atomic resolution. This means that the resolution is actually so high that it is possible to see and distinguish the individual atoms (0.2 nm = 2 * 10-10 m) on the surface. The invention of the STM earned Binning and his co-worker Heinrich Rohrer (at IBM Zürich) the Nobel Prize in Physics in 1986.

Scanning Tunnelling Microscope: The STM is a fundamental tool in nanoscience and nanotechnologies. It is used in both industrial and fundamental research to obtain atomic-scale images of metal and semiconducting surfaces. It provides a three-dimensional profile of the surface roughness, allowing the observation of surface defects and the determination of the size and conformation of molecules and aggregates. Another astonishing property of the STM is that it can be used to manipulate or move individual atoms, trigger chemical reactions, as well as performing electronic spectroscopy. If it is necessary to scan a surface which in itself is not electrically conductive, it can be coated with a very thin layer of a conductive material such as gold. This does, however, imply that the STM is less suitable for some studies such as DNA which is not conductive. For these types of samples, other SPM techniques are more suitable, such as the Atomic Force

Microscope (AFM). An AFM does not measure the tunnel current, but the forces between the tip and the surface and, therefore, does not require the surface to be conductive. The AFM was developed in 1985, also by Binning and co-workers at IBM Zürich. It was developed specifically to image materials that are insulating.

Atomic Force Microscope: The Atomic Force Microscope (AFM) was developed specifically to overcome the intrinsic limitations of the STM, which is not suitable for imaging surfaces coated with biological entities such as DNA or proteins. The AFM operates in air and not under a vacuum. Some versions of the instrument also allow operation in liquid, which is very advantageous when imaging biological samples that often need buffers to remain biologically active. The AFM measures the interaction force, both attractive or repulsive, between the probe and the surface. The probe is continuously moved along the surface and the cantilever detection is constantly monitored. A feedback loop continuously changes the height of the probe on the surface in order to keep the applied force constant. The vertical movement of the probe is recorded to create a topographic map of the surface under study.

Spectroscopy Method: Spectroscopy is defined as the branch of science that is concerned with the investigation and measurement of spectra produced when matter interacts with or emits electromagnetic (EM) radiation. Depending on the wavelength of the electromagnetic radiation used and the type of interaction with matter that occurs (absorption, scattering, etc.), different spectra are measured from which much information can be inferred. Some relevant methods are briefly discussed.

X-ray Method: X-ray methods involve exciting a sample either with X-rays or with electrons. The energy of emitted X-rays is equal to the difference between the binding energies of the electrons involved in the transition. In the context of nanomaterials, the most important method is small-angle X-ray scattering (SAXS) and can analyse particle sizes of the order of 1–100 nm. The method can be used to image powders in the dry state or suspended in a medium. The method can also be used to measure the nanoparticle size.

Surface-enhanced Raman Scattering: Metal surfaces with nanometre scale roughness have In simple terms, Raman scattering is the inelastic scattering of photons. Normally, when light is scattered from an atom or molecule, it has the same energy and wavelength as the incident light. This is an elastic scattering. However, a small fraction of the scattered light (approximately 1 in 10 million photons) is scattered by excitation, with the scattered photons having energy different to the frequency of the incident photons. Metal surfaces with nanoscale roughness increase the Raman scattering of molecules absorbed on them. This effect is due to chemical and electromagnetic factors, as well as increased surface area. The surface enhanced Raman scattering (SERS) effect can induce a signal enhancement of up to 108 times. In one specific case, it has been possible to achieve a Raman enhancement effect of 1015 times! This means that the SERS effect makes it possible to push the detection limit of surface detection techniques. The SERS signal depends on the characteristics of the nano-substrate: the size, shape, orientation and composition of the surface nano-roughness. Advancements in SERS technology will allow detection at the attomole (10-18 mol) level, and single molecule detection.

4.7 APPLICATIONS OF NANOTECHNOLOGY

Nanoscale materials, as seen above, have been used for decades in several applications, in a wide range of products. Among the most well known are a glass for windows which is coated with titanium oxide nanoparticles that react to sunlight to break down dirt. When water hits the glass, it spreads evenly over the surface, instead of forming droplets, and runs off rapidly, taking the dirt with it. Nanotechnologies are used by the car industry to reinforce certain properties of car bumpers and to improve the adhesive properties of paints. Some uses of nanotechnologies are as follows:

Nanorobotics: An emerging application is creating robots whose components are at or close to the scale of a nanometer (10^{-9} meters). Nanorobotics refers to the nanotechnology engineering application of designing and building nanorobots, with devices ranging in size from 0.1–10 micrometers and constructed of nanoscale or molecular components. The names nanobots, nanoids, nanites, nanomachines, or nanomites have also been used to describe these devices. Another definition is a robot which allows precision interactions with nanoscale objects or can manipulate in nanoscale resolution. Such devices are more related to microscopy or scanning probe microscopy. Following the microscopy definition, even a large apparatus such as an atomic force microscope can be considered a nanorobotic instrument when configured to perform nano-manipulation. Albert Hibbs, a graduate student of Richard Feyman, had originally suggested the idea of medical use for Feynman's theoretical micromachines. Hibbs suggested that certain repair machines might one day be reduced in size to the point that it would, in theory, be possible to "swallow the doctor". The idea was incorporated into Feynman's 1959 essay "There's Plenty of Room at the Bottom". Basic idea was that a robot can perform certain functions that humans cannot, like a microscopic robot performing microscopic tasks. Nanorobots find applications in various fields. Widely used in the field of Space technology and Medicine including heart surgeries, cancer treatment, dentistry etc. In the military, they are used as an improved body armor capable of self-repair if damaged. Self Replicating Nanorobots will be used to treat injured people and damaged equipment on the battlefield and as eavesdropping devices that are practically undetectable. Due to its small size and ease in propelling and placing in orbit in less time, nanorobots finds considerable application in space technology. These bots now can be used in carrying out construction projects in hostile environments. Just a handful of self replicating robots, utilizing local materials and local energy can completely construct space habitat by remote control so that the inhabitants need only show up with their suitcases. An engineer or group of engineers could check up on the construction of the habitats via telepresents utilizing cameras and sensors from the comfortable confines of the earth.

Nubot: It is an abbreviation for "nucleic acid robot." They are organic molecular machines at the nanoscale and are also known as DNA machine. DNA structure can provide means to assemble 2D and 3D nanomechanical devices. These machines can be activated by using small molecules, proteins and other molecules of DNA. Biological circuit gates based on DNA materials have been engineered as molecular machines to allow in-vitro drug delivery for targeted health problems. Bacteria based approach uses biological microorganisms, like the bacterium Escherichia coli for delivering foreign DNA into tumor cells. Electromagnetic fields normally control the motion of this kind of biological integrated device. Chemists at the University of Nebraska have created a humidity gauge by fusing a bacterium to a silicone computer chip. There are virus based systems to attach to cells and replace DNA. They go through a process called reverse transcription to deliver genetic packaging in a vector. These Gene Therapy vectors have been used in cats to send genes into the genetic modified animal "GMO" causing it display the trait

Fractal Robot: Fractal robot is a new kind of robot made from motorized cubic bricks that move under computer control. These cubic motorized bricks can be programmed to move and shuffle themselves to change shape to make objects likes a house potentially in a few seconds because of their motorized internal mechanisms.

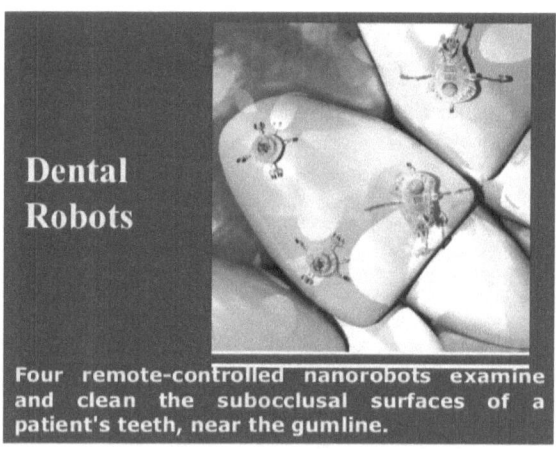

Four remote-controlled nanorobots examine and clean the subocclusal surfaces of a patient's teeth, near the gumline.

Dental Nanorobots: They examine and clean the subocclusal surfaces of a patient's teeth, near the gumline. A mouthwash full of smart nanomachines could identify and destroy pathogenic bacteria while allowing the harmless flora of the mouth to flourish in a healthy ecosystem.

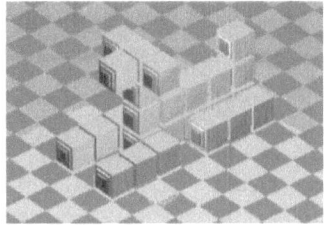

Space Robot: Swarms are Nanorobots for use in space. They theoretically will be akin to flexible cloth like material and being composed of what is called Bucky tubes, this cloth will be as strong as diamond. By addition of a nano computer to this nano machine, a smart cloth is formed. This smart cloth can then be used to keep astronauts from bouncing around inside their spacecraft while they sleep, a problem that arises when the auto pilot computer fires the course correction rockets. The cloth like material will be able to offset the sudden movements and slowly move the astronaut back into position. Nanomaterials make lightweight solar sails and a cable for the space elevator possible. By significantly reducing the amount of rocket fuel required, these advances could lower the cost of reaching orbit and traveling in space. Radiation shielding is the area where Nanotechnology makes a major contribution to human space flight. NASA says that the risks of exposure to space radiation are the most significant factor limiting humans' ability to participate in long duration space missions. Space radiation is qualitatively different from the radiations that humans encounter on the earth's surface. Once the astronauts leave the earth's protective magnetic field and atmosphere, they become exposed to ionizing radiation in the form of charged atomic particles traveling at close to the speed of light. Highly charged, high energy particles pose risk to humans in space. A long-term exposure to this radiation may lead to DNA damage and cancer. To protect their human cargo, spacecrafts will need special shields incorporating materials consisting of lighter elements such as hydrogen, Boron and Lithium. However, extra shielding comes at a significant price in the form of extra weight, more fuel and increased flight costs. Nanotubes materials made from carbon nanotubes can be employed to reduce the weight of spaceships or in increasing the structural strength. The carbon nanotubes may also be used to make the cable needed for the space elevator, a system that could reduce the cost of sending material into orbit.

Nanomedicine: Nanomedicine is the medical application of nanotechnology and ranges from the medical applications of nanomaterials, to nanoelectronic biosensors, and even future applications of molecular nanotechnology. The emerging discipline of nanomedicine brings nanotechnology and medicine together in order to develop novel therapies and improve existing treatments. In nanomedicine, atoms and molecules are manipulated to produce nanostructures of the same size as biomolecules for interaction with human cells. This procedure offers a range of new solutions for diagnoses and smart treatments by stimulating the body's own repair mechanisms. It will enhance the early diagnosis and treatment of diseases such as cancer, diabetes, Alzheimer's, Parkinson's and cardiovascular diseases. Preventive medicine may then become a reality.

Drug Delivery: Therapy on the basis of externally administered (oral, intravenous, subcutaneous, etc.) pharmaceutical preparations suffers from three severe problems: (i) as a rule, they have had to be administered systemically, and since their therapeutic effect is usually a differential one (i.e., it is slightly more toxic to the offending cells than to the rest of the body) the entire body is poisoned to some degree; (ii) many therapeutically efficacious compounds are destroyed by the regular defence mechanisms of the body against foreign invaders before they arrive at their destination; and (iii) the bilayer lipid membrane, ubiquitous in eucaryotic cells, is increasingly recognized as an important target for medicinal drugs, but the hydrophobicity of the drug molecule required to ensure its high affinity for the membrane is incompatible with transporting it through the predominantly aqueous media between the point of administration and the target membrane. These are of course generalizations with exceptions, but numerous examples show their widespread validity. Nucleic acids might be very valuable, and certainly they could be targeted very specifically, if they could penetrate the nuclei of cells, but they are hydrolysed and destroyed by circulating enzymes long before. The aim of drug delivery is to encapsulate a therapeutic agent to disguise its properties until it reaches its target and is released. The encapsulation must therefore be able to respond to its environment. There is a general requirement that it should not excite an adverse immune response; that is, its surface must be biocompatible, which typically means it should not bind and denature the proteins circulating in the blood. A simple example is to place a drug destined for the stomach in a hollow sphere of calcium carbonate. This material will dissolve in the high concentration of hydrochloric acid found in the stomach, releasing the drug. In all other respects it should be possible to handle the encapsulated drug in the same way as its unencapsulated congener, which is where the nanotechnology comes in: a nanosphere can be treated in the same way as a largish molecule. A benefit of using nanoscale for medical technologies is that smaller devices are less invasive and can possibly be implanted inside the body besides having lower biochemical reaction times. These devices are faster and more sensitive than typical drug delivery. As explained above, efficacy of drug delivery through nanomedicine is largely based upon, efficient encapsulation of the drugs, successful delivery of drug to the targeted region of the body, and successful release of the drug. Drug delivery systems, lipid or polymer based nanoparticles, can be designed to improve the pharmacokinetics and biodistribution of the drug. Complex drug delivery mechanisms are being developed, including the ability to get drugs through cell membranes and into cell cytoplasm. Drug delivery systems may also be able to prevent tissue damage through regulated drug release; reduce drug clearance rates; or lower the volume of distribution and reduce the effect on non-target tissue. However, the biodistribution of these nanoparticles is still imperfect due to the complex host's reactions to nano and microsized materials and the difficulty in targeting specific organs in the body. Nevertheless, a lot of work is still ongoing to optimize and better understand the potential and limitations of nanoparticulate systems.

Biosensors: The main point of clinical biosensing is continuous, non-invasive monitoring. Currently, most tests require a sample of the relevant biofluid (e.g., blood) to be drawn from the patient. For most people this is a somewhat unpleasant procedure. It is, however, recognized that much more insight into a patient's pathological state could be obtained by frequent, and continuous, monitoring. At present, this is only possible in intensive care unit. It is a difficult technical problem to extend non-invasive analysis to the plethora of biomarkers currently under intensive study as symptomatic of disease or incipient disease. An alternative approach is to develop sensors so tiny that they can be semi-permanently implanted inside the body, where they can continuously monitor their surroundings. Because of the large and growing number of afflicted people, diabetes has received overwhelming attention. The sensing requirement is for glucose in the blood. The glucose sensor follows the classic biosensing design: a recognition element to capture the analyte (glucose) mounted on a transducer that converts the presence of captured analyte into an electrical signal. The recognition element is typically a biological molecule, the enzyme glucose oxidase, hence this device can be categorized as both nanobiotechnology.

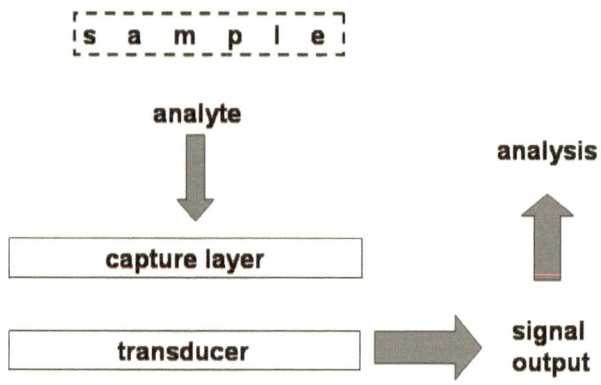

Tissue Engineering: Nanotechnology may be used as part of tissue engineering to help reproduce or repair damaged tissue using suitable nanomaterial based scaffolds and growth factors. Tissue engineering if successful may replace conventional treatments like organ transplants or artificial implants. Nanoparticles such as graphene, carbon nanotubes, molybdenum disulfide and tungsten disulfide are being used as reinforcing agents to fabricate mechanically strong biodegradable polymeric nanocomposites for bone tissue engineering applications. The addition of these nanoparticles in the polymer matrix at low concentrations leads to significant improvements in the compressive and flexural mechanical properties of polymeric nanocomposites. Potentially, these nanocomposites may be used as a novel, mechanically strong, light weight composite as bone implants. For example, a flesh welder was demonstrated to fuse two pieces of chicken meat into a single piece using a suspension of gold-coated nanoshells activated by an infrared laser. This could be used to weld arteries during surgery. Another example is nanonephrology, the use of nanomedicine on the kidney.

Medical Devices: Neuro-electronic interfacing is a visionary goal dealing with the construction of nanodevices that will permit computers to be joined and linked to the nervous system. This idea requires the building of a molecular structure that will permit control and detection of nerve impulses by an external computer. A refuelable strategy implies energy is refilled continuously or periodically with external sonic, chemical, tethered, magnetic, or biological electrical sources, while a non-refuelable strategy implies that all power is drawn from internal energy storage which would stop when all energy is drained. A nanoscale enzymatic biofuel cell for self-powered nanodevices have been developed that uses glucose from biofluids including human blood and watermelons. One limitation to this innovation is the fact that electrical interference or leakage or overheating from power consumption is possible. The wiring of the structure is extremely difficult because they must be positioned precisely in the nervous system. The structures that will provide the interface must also be compatible with the body's immune system.

Molecular Nanotechnology: It is a futuristic subfield of nanotechnology regarding the possibility of engineering molecular assemblers, machines which could re-order matter at a molecular or atomic scale. Nanomedicine would make use of these nanorobots, introduced into the body, to repair or detect damages and infections.

Nanosurgery: Nano-robots that can function as tiny surgeons are no more in the realm of science fiction. Extensive research is being carried out into various types of computer and robot assisted surgery and virtual reality interfaces. Surgical nano robots could be introduced into the body through the vascular system or at the ends of catheters into various vessels and other cavities in the body. A surgical nanorobot, programmed or guided by a human surgeon, could act as a semi-autonomous on site surgeon inside the body. Such a device could perform various functions such as searching for pathology and then diagnosing and correcting lesion by nano-manipulation, coordinated by an onboard computer while maintaining contact with the supervising surgeon via coded ultrasound signals. Biocompatible surgical nanorobots can find and eliminate isolated cancerous cells, remove micro vascular obstructions and recondition vascular endothelial cells. They can perform noninvasive tissue and organ transplants, conduct molecular repairs on traumatized extracellular and intracellular structures and even exchange new whole chromosomes for old ones inside individual living cells. The greatest power of nanotechnology is expected in the

2020s, when complete artificial nanorobots using rigid diamondoid nanometer scale parts like molecular gears and bearings may be designed. These nanorobots will possess full panoply of autonomous subsystems including onboard sensors, motors, manipulators, power supplies, and molecular computers. Spontaneous self assembly of all these nanoscale components in the right sequence will be a challenging task.

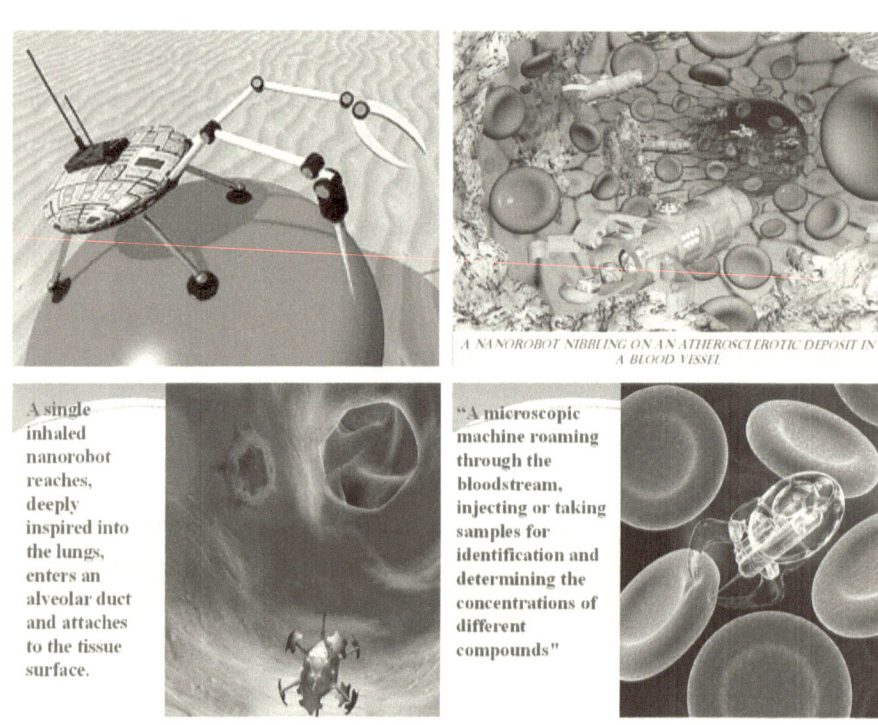

A NANOROBOT NIBBLING ON AN ATHEROSCLEROTIC DEPOSIT IN A BLOOD VESSEL

A single inhaled nanorobot reaches, deeply inspired into the lungs, enters an alveolar duct and attaches to the tissue surface.

"A microscopic machine roaming through the bloodstream, injecting or taking samples for identification and determining the concentrations of different compounds"

We all know that cardiac diseases are major cause of mortality, and morbidity in human beings. Currently, nanotechnology offers a broad platform in the field of cardiovascular science by offering tools to explore the frontiers of cardiac science at the cellular level. Nanotechnology-based tools can be effectively used to treat the cardiovascular diseases. These machines have three key elements meant for sensing, decision making, and carrying out the intended purpose.

Sun Glass: Nanotechnology offers scratch-resistant coatings based on nanocomposites that are transparent, ultra-thin, simple to care for, well-suited for daily use and reasonably priced.

Textile: Nanotechnology is incorporated to make practical improvements to properties such as wind proofing, waterproofing, preventing wrinkles or stains, and guarding against electrostatic discharges. The windproof and waterproof properties of ski jacket, for example, are obtained not by a surface coating of the jacket but by the use of nanofibers. Future projects include clothes with additional electronic functionalities, so-called "smart clothes" or "wearable electronics". These could include sensors to monitor body functions or release drugs in the required amounts, self-repairing mechanisms or access to the Internet.

Sports Equipment: These manufacturers are also turning to nanotech. A high-performance ski wax, which produces a hard and fast gliding surface, is already in use. The ultra thin coating lasts longer than conventional waxing systems. Tennis rackets with carbon nanotubes have increased torsion and flex resistance. The rackets are more rigid than current carbon rackets and pack more power. Long-lasting tennis-balls are made by coating the inner core with clay polymer nanocomposites and have twice the lifetime of conventional balls.

Sunscreens and Cosmetics: Nanotech based cosmetics are already widely used. Customers like products that are translucent because they suggest purity and cleanliness, and cosmetic companies have discovered that when lotions are ground down to 50 or 60 nms, they let light through. For sunscreens, mineral nanoparticles such as titanium dioxide offer several advantages. Traditional chemical UV protection suffers from its poor long-term stability. Titanium dioxide nanoparticles have a comparable UV protection property as the bulk material, but lose the cosmetically undesirable whitening as the particle size is decreased. For anti-wrinkle creams, a polymer capsule is used to transport active agents like vitamins.

Nanoseeds: In Thailand, scientists at Chiang Mai University's nuclear physics laboratory have rearranged the DNA of rice by drilling a nano-sized hole through the rice cell's wall and membrane and inserting a nitrogen atom. So far, they've been able to change the colour of the grain, from purple to green.

Energy: Breakthroughs in nanotechnology could provide technologies that would contribute to world-wide energy security and supply. Numerous areas have been identified in which nanotechnology could contribute to more efficient, inexpensive, and environmentally sound technologies than are presently available. The most significant contributions may be to applications such as better materials for exploration equipment used in the oil and gas industry, solar power; wind; clean coal; fusion reactors; new generation fission reactors; fuel cells; batteries; hydrogen production, storage and transportation; and a new electrical grid that ties all the power sources together. Bell Labs is exploring the possibility of producing a microbattery that would work 20 years by postponing the chemical reactions that degrades traditional batteries. The battery is based on discovery that liquid droplets of electrolyte will stay in a dormant state atop microscopic structures called "nanograss" until stimulated to flow, thereby triggering a reaction producing electricity. Other researchers hope to dispense with batteries completely by developing nanotubes-based "ultra" capacitors powerful enough to propel hybrid-electric cars. Photovoltaics is another area where nanotech is already providing products that could have a significant impact. Three US-based solar cell start-ups viz Nanosolar, Nanosys and Konarka Technologies, and corporate players including Matsushita and STMicroelectronics are striving to produce photon-harvesting materials at lower costs and in higher volumes than traditional crystalline silicon photovoltaic cells. Nanosolar has developed a material of metal oxide nanowires that can be sprayed as a liquid onto a plastic substrate where it self- assembles into a photovoltaic film. Nanosys intends its solar coatings to be sprayed onto roofing tiles. And Konarka is developing plastic sheets embedded with titanium dioxide nanocrystals coated with light-absorbing dyes. If nanotech solar fabrics could be applied to, e.g., buildings and bridges, the energy landscape could change in important ways. Integrated into the roof of a bus or truck, they could split water via electrolysis and generate hydrogen to run a fuel cell.

Agriculture: Researchers are developing a range of inexpensive nanotech applications to increase soil fertility and crop production, and help eliminate malnutrition which contributes to more than half the deaths of children under five in developing countries. Nanotech materials are in development for the slow release and efficient dosage of fertilisers for plants and of nutrients and medicines for livestock. Other agricultural developments include nanosensors to monitor the health of crops and farm animals and magnetic nanoparticles to remove soil contaminants.

Nanosoftware: Some of computer programs used for modeling of nanostructures at classical and quantum levels are as follows:

- Ascalaph Designer
- Atomistix ToolKit and Virtual NanoLab
- CST STUDIO SUITE(TM)
- CoNTub
- Deneb Powerful Graphical User Interface (GUI) for SIESTA, VASP, QE, etc., DFT calculation packages
- Nanohub allows simulating geometry, electronic properties and electrical transport phenomena in various nanostructures
- Ninithi-A Carbon Nanotube, Graphene and Fullerene modelling software
- Nanoengineer-1 (developed by company Nanorex, but the website doesn't work, probably no more available)
- NEMO 3D enables multi-million atom electronic structure simulations in empirical tight binding. It is open source. An educational version is on nanoHUB as well as Quantum Dot Lab
- Nanotube Modeller
- Materials Design MedeA
- Materials Studio
- MD-kMC
- SCIGRESS
- TubeASP
- Tubegen
- Wrapping

4.8 TWO CONCEPTS

Nanification: The word nanification, makes us think of miniaturization but in a wholesome way. To nanify electronics, for example, is not only to make individual components smaller (right down to the nanoscale) but also to adapt all parts of the industry to that situation, including design aspects. In short, nanification means introducing nanotechnology in an integrated rather than a piecemeal fashion. Hence, to nanify manufacture is, ultimately, to introduce molecular manufacturing, which involves not only the actual assembly devices themselves, but also logistics, indeed the entire supply chain, and the reorganization of the economic system.

Sensorization: It literally means incorporation of vast numbers of sensors, something that is only feasible if they are nanoscale sensors, from the viewpoints of both cost and space requirements. Sensorization is likely to lead to a qualitatively different way of handling situations in at least four areas:

- Structural engineering: Sensors will be incorporated throughout the structure, like embedded in concrete, or in the wings of an aircraft. The output of these sensors is indicative of strain, the penetration of moisture, and so forth.
- Process engineering: Sensors embedded throughout machinery and reaction vessels will monitor physical parameters like temperature and chemical variables like concentration variable of a selected substance.
- Biosensors will be incorporated into the human body to continuously monitoring physiological variables.

Sensors will be dispersed throughout the environment such as along rivers and in lakes, reporting the purity of water, and so forth. The concept is in some ways akin to what is already taking place in microfarming; wherein intervention is guided by high-resolution satellite images of fields, indicating local moisture.

4.9 NANOTECHNOLOGY INITIATIVE OF GOVT. OF INDIA

The Nano Mission is a Mission-Mode programme within Department of Science & Technology. At the apex level, it is steered by Nano Mission Council (NMC). It is an umbrella programme for capacity building which envisages the overall development of this field of research in the country and to tap some of its applied potential for nation's development. The technical programmes of the Nano Mission are also being guided by two advisory groups, viz. the Nano Science Advisory Group (NSAG) and the Nano Applications and Technology Advisory Group (NATAG). International Collaborative Programmes in Nano Science and Technology has prominently figured in all S&T cooperation agreements entered into in recent times. Joint R&D activities are already taking place with several countries. For example, with the US, several projects have been funded in composites, nano-encapsulating materials, etc. Several Indo-US Workshops have also been held. With Germany, a programme on engineered functional nanocomposites has started which would focus on magnetic properties, magnetic interactions, gas-solid interactions including catalysis, etc. Programmes are also on with Italy, EU and developing with Taiwan. ARCI, Hyderabad, which is an autonomous institute of DST has active programmes in nanomaterials with institutions in Russia, Ukraine, Japan, Germany and USA. In order to focus expertise in research towards developing products and processes of direct interest to industries, DST, under the Nano Programme, has promoted Joint Institution-Industry Linked Projects and some other Public Private Partnership activities. These activities will help us to simultaneously leverage the scientific knowledge base existing in our research and educational institutions and the commercial vision of our industry to generate competitive technologies leading to products and devices.

Eleven Units/Core Groups on Nano Science have been sanctioned across the country as centers of excellence which house some of the sophisticated facilities for sharing with other scientists in the region and would help in promoting scientific research on nanoscale systems in a decentralized fashion. Seven Centers for Nano Technology focusing on development of specific applications have also been established. In addition, a center of excellence on Computational Materials Science has also been established at JNCASR, Bangalore.

Unit of Nano Science		
Sl. No.	Unit of Nano Science	Programme Coordinator
1.	IIT Madras, Chennai	Prof. T. Pradeep
2.	IACS, Kolkata	Prof. D. Chakravorty
3.	University of Pune	Prof. (Mrs.) S.K. Kulkarni
4.	S.N. Bose National Centre for Basic Sciences, Kolkata	Prof. A.K. Raychaudhuri
5.	NCL, Pune	Dr. Sivaram
6.	JNCASR, Bangalore	Prof. G.U. Kulkarni
7.	BHU, Varanasi	Prof. O.N. Srivastava
8	IIT Kanpur, Kanpur	Prof. Ashutosh Sharma
9	IISc, Bangalore	Prof. S. Chandrasekaran
10	IIT Delhi, New Delhi	Prof. B.R. Mehta
11.	SINP, Kolkata	Prof. M.K. Sanyal

Centre for Nano Technology		
Sl. No.	Centre for Nano Technology	Programme Coordinator
1.	Amrita Institute of Medical Sciences, Kochi, Kerala (Implants, Tissue Engineering, Stem Cell Research)	Dr. Shantikumar V Nair, Biomedical Engineering Centre, Amrita Institute of Medical Sciences, Elamakkara PO, Koshi, Kerala
2.	S.N. Bose National Centre for Basic Sciences, Kolkata (NEMS & MEMS / Nano products)	Prof. A.K. Raychaudhuri, S.N. Bose National Centre for Basic Sciences, J.D. block, Sector III, Salt Lake, Kolkata
3.	Tata Institute of Fundamental Research (Nanoscale phenomena in biological systems & materials)	Dr. G.V. Shivshankar National Centre for Biological Sciences, TIFR, GKVK Campus, Bellary Road, Bangalore
4.	IIT-Bombay, Mumbai (Nanoelectronics, polymer nanosensors, nanobiotechnology)	Prof. Ashok Misra Director, IIT-Bombay Powai, Mumbai
5.	Indian Institute of Science, Bangalore (Nanodevices, Nanocomposites, Nanobiosensors)	Prof. S. Chandrasekaran Division of Chemical Sciences Indian Institute of Science Bangalore
6.	IIT, Kanpur (Printable Electronics, Nanopatterning)	Prof. Y.N. Mohapatra Deptt. of Physics, IIT-Kanpur Kanpur – 208016
7.	Indian Association for the Cultivation of Science (Photovoltaics & Sensor Devices)	Prof. D.D. Sarma Centre for Advanced Materials, Indian Association for the Cultivation of Science, Kolkata

Centre for Computational Materials Science		
Sl. No.	Centre for Computational Materials Science	Programme Coordinator
1.	Centre for Computational Materials Science at Jawaharlal Nehru Centre for Advanced Scientific Research, Bangalore	Prof. Balasubramanian Sundaram, JNCASR, Jakkur PO, Bangalore

Major Research Centers and Publications:

Name	h-index	Number of Articles
Council of Scientific and Industrial Research (2016)	13	1543
Indian Institute of Technology Kharagpur (2016)	8	361
Indian Institute of Technology Madras (2016)	7	278
Indian Institute of Technology Delhi (2016)	7	259

Name	h-index	Number of Articles
VIT University (2016)	8	259
Indian Institute of Technology Bombay (2016)	6	257
Indian Institute of Technology Kanpur (2016)	8	256
Indian Institute of Technology Roorkee (2016)	11	255
Jadavpur University (2016)	6	220
Banaras Hindu University (2016)	6	210

4.10 NANOTECHNOLOGY COMPANIES IN INDIA

Adnano Technologies: Adnano Technologies is a supplier of various forms of graphene and multiwalled carbon nanotubes. They also provide analytical services like FESEM, TEM, AFM, FTIR, XRD, XPS, Contact Angle, BET, Zeta sizer and Master sizer.

Advanced Nano Tech Lab: A manufacturer of coating products.

Auto Fibre Craft: AFC Powders is a company involved in manufacturing specialized nanomaterials. Currently it is manufacturing Nano-size Silver Powder for use in electronic applications for e.g. making conductive inks and pastes, RFID. This product is RoHS compliant (restriction of hazardous substance)

AVANSA Technology & Services: This company specializes in analytical characterization, consultancy, and synthesis of nanomaterials serving to nanotechnology-based industries, universities and institutes. They also manufacture carbon nanotubes, graphene, and various nanoparticles.

Bee Chems: A chemicals company focused on the silica and alumina industries. Manufactures various grades of Nano Silica products.

Bilcare: Bilcare has developed a unique security technology called nonClonable. The technology innovatively exploits the intrinsic nature of nano and micro-structured composites together with their magnetic and optical properties to provide a foolproof security system.

Bottom Up Technology Corporation: Manufactures graphene and caron nanotubes.

Dabur Pharma: One of the company's delivery systems is in the most advanced stages of clinical development for a novel drug delivery system for Paclitaxel. Because of the better safety and pharmacokinetic profile, the polymeric nanoparticle delivery system is seen as a potential super generic drug.

Egoma Technologies: Specializes in producing customizable solutions to Ball Mill requirements and nanopowders. They also provide consultancy in material related industrial and research R&D related problems.

Eris Technologies: Eris Technologies is a software development company that also provided industrial training in areas like nanotechnology. Certification in nanotechnology covers course in applied nanotechnology where the dimensions and tolerances in the range of. 1 nm to 100 nm play a critical role. The course covers nanotechnology basics, manufacturing process, lithography, CNT, nanocomputers, nanomedicine and nanodiamond.

Icon Analytical Equipment: Is a distributor of analytical instruments, with a focus on nanotechnology and related analytical techniques.

Kerala Minerals & Metals (KMML): Manufactures various grades of titanium dioxide nanoparticles.

Meda Biotech: is a clinical-stage biopharmaceutical company developing a new class of Hybrid- Nanoengineering™ platform enables the design, engineering and manufacturing of Hybrid water soluble drugs with unprecedented control over drug properties to maximize trafficking to disease sites, dramatically enhancing efficacy while minimizing toxicities.

Micromaterials (India): is a Bangalore based company focused on developing innovative nano and micro technologies and materials catalysts. The new generation catalysts are the result of a radically new patented process.

Mittal Enterprises: Mittal Enterprises has distinction in manufacturing Nanofluid Interferometer and other Laboratory and Scientific Instruments covered with Trademark, Design Registration and Copyright. Established in 1976, they are an ISO 9001:2008 certified company.

Nano Cutting Edge Technology Nano CET: Contract Research and New Product Development Work involving biostabilized nanoparticle technology.

Nanomics Technologies: The company manufactures a wide range of nanomaterials such as carbon nanotubes, graphene, nanoalloys, nanowires, and nanoparticle powders and suspensions.

Nano Research Elements: A provider of nanomaterials.

Nanoshel: Nanoshel makes more than 50 types of nanomaterials, among which the main products are nanotubes, SWCNT's, MWCNT's, nanoparticles.

Nano Sniff Technologies: Nano Sniff is a commercial spin-off from the Center of Excellence in Nanoelectronics (CEN) at IIT Bombay and was formed to productize various technologies developed as a part of research work conducted at CEN. NanoSniff is the first Indian company to successfully commercialize to microcantilever and microheater sensor technologies and sell these devices off-the-shelf.

Nanospan: The company is active in manufacture, supply and application of graphene related materials. They offer a range of graphene types, functionalized graphene, graphene intermediates, carbon nanotubes and nanomaterials. They provide analytical testing & engineering services for emerging nanotechnology-based products in the areas of energy storage, electronics, polymer/resin/lubricants, electronic inks and 3D printing. They also provide nanomaterial characterization &testing services like HR-TEM, FESEM, FTIR, XRD, BET, Raman Spectroscope, AFM.

NanoXpert Technologies: KCIL-NanoXpert Technologies is an intrapreneurial arm of KCIL (Kairav Chemofarbe Industries Ltd) India for high-tech nanoparticle business, it has a technological collaboration with National Chemical Laboratories, Pune, India, under the prestigious Prime Minister?s Fellowship Program. NanoXpert's products have been developed by experts in the field with individual experience of over a decade.

Navran Advanced Nanoproducts Development: Develops and manufactures nanoproducts such as polymerized toners.

Neo-Ecosystems: is specialized on researching and production of metal nanopowders.

Nilima Nanotechnologies: Nilima Nanotechnologies offers a wide range of nanotechnology based coatings with protective properties for various surfaces.

Platonic Nanotech: The company uses its proprietary bottom-up process for the production of high quality graphene.

Quantum Corporation: Quantum Corporation (QCorp) is the parent company of group of companies, head quartered in Bangalore, India. QCorp was established in 2007 with a vision to create world class Nanomaterials and Nanocomposites with strong Intellectual Property that are changing the properties of products across the globe. QCorp has developed high quality Smart Polymers, Nanomaterials and Nanocomposites as core materials for manufacturers in

Telecommunications, Electronics, Drug Delivery, Conductive films, Lighting and Energy industries – without the need to change their existing processes.

Reinste Nano Ventures: A manufacturer of nanomaterials.

Saint-Gobain Glass: The company manufactures SGG NANO, a high performance coated glass with advanced energy efficient solar control and thermal insulation (low e) properties. This Advanced Solar Control and Thermal Insulation (low e) Glass is manufactured by deposition of multiple layers of highly specialized nano-metric metallic oxides / nitrides by a process of magnetically enhanced nanotechnology-based cathodic sputtering under vacuum conditions.

Sisco Research Laboratories (SRL): The company is a manufacturer and supplier of about 3500 Laboratory Chemical and Allied Products. They specialize in Molecular Biology Reagents, Biochemicals, Enzymes, Nanopowders and Carbon Nanotubes, Organic & Inorganic Intermediates, High Purity Solvents, Culture Media &Bio Lit™ DNA & Protein Tools for PCR and regular Molecular Biology.

Smart Nanoz: The company manufactures nanoparticles.

Ultrananotech: Manufactures nanoparticles and graphene.

United Nanotech Innovations: A manufacturer of graphenes and nanocomposites.

Velbionanotech: An R&D company in the nanobio and nanomedicine area.

Few technologies developed successfully in the field of Nanotechnology are enumerated here.

Health Care – Silicon Locket: A web enabled efficient and intelligent remote health care system for cardiac diagnostics entitled 'Silicon Locket' has been developed.

MEMS Pressure Sensor: Design, fabrication and testing of a Nanocrystalline silicon MEMS pressure sensor in the range of 1 mbar to1 bar.

Silver Nano Powder: A process has been developed for the generation of Silver Nano powder at 200 gm batch level.

Ge Quantum Dots: A technology for growing size and shape controlled Ge quantum dots with excellent optical properties has been developed.

Synthesis of Nanoparticles: Silicon nanoparticles have been synthesized on silicon substrate with different insulating layers namely SiO_2/Si_3N_4, by using two different processes. These include Low Pressure Chemical Vapor Deposition (LPCVD) method and Plasma Enhanced Chemical Vapor Deposition (PECVD).

Nano Measurement: Software developed for calculating nano measurement uncertainty.

Polymeric Cantilever Technology: Polymeric cantilever technology with an embedded polysilicon piezo-resistor has been demonstrated.

Explosive Vapour Sensor: Explosive Vapour Sensor using poly (3-hexythiophene) and Cu-tetraphenyl-porphyrin composite based organic field effect transistor has been developed. This technology can detect vapours of explosives.

Sensor Platform: A low cost sensor platform for determining ionizing radiation using sensors based on organic semiconducting materials has been demonstrated.

Micro-stereo-lithography: For Polymer & Ceramics: Micro-Stereo-lithography for generating shapes polymer & ceramics has been developed.

Technology for the development of low cost AFM has been developed.

The following patents and copyright have been filed:

- Novel Ways of Introducing High Voltage Handling Capabilities in FinFET Technologies – US patent.
- Sandwich Tunneling Barrier FET – US patent.
- Dual Gate STI DeMOS for improved mixed signal and hot carrier behavior – US patent.
- Performance Improvement of Tunnel FET Devices using Halo-Doping, Graded Silicon-Germanium, Schottky Junction and New Device Structures – US patent.
- Process for Novel Multiferroic based Micro Electromechanical Systems (MEMS) Actuators Functioning at Room Temperature – Indian Patent.
- Low-Energy Successive Approximation Register Analog to Digital Converter Technique and Circuit – Indian patent.
- Nonvolatile Floating Gate Analog Memory Cell – Indian Patent.
- An interface circuit between a sensor and a signal conditioning circuit – Indian patent.

- Single Halo DeMOS for robust protection in advanced high voltage CMOS – US patent application.
- A DeMOS Device realized using dual STI process – US patent.
- Microheater based explosive sensor – Indian patent.
- Independently Driven Double Gate (IDDG) nonvolatile floating gate analog memory cell – Indian patent.
- Method and device for determining ionizing radiation – Indian patent.
- A novel dry method of surface modification of SU8 for immobilization of biomolecules using hotwire pyrolytic process – Indian patent.
- Method for doping a fin based semiconductor device – US patent.
- Integrated pressure and temperature sensor based on porous silicon – Indian patent.
- An IGBT device with plugged-in SCR for robust ESD protection in FinFET technology – US patent.
- A Novel Architecture for Improving Slew Rate in FinFET based Op-Amp and OTAs – US patent.
- Software entitled "Software for uncertainty in measurement using LabView (NPLI-SUM)" – Indian copyright.
- An N-Type Tunnel-FET device with strained SiGe layer at source – Indian Patent.
- A sub-threshold forced plate FET sensor for sensing inertial displacement, a method and system thereof – Indian Patent.
- A sub-threshold elastic deflection FET sensor for sensing pressure/force, a method and system thereof – Indian Patent.
- A sub-threshold CapFET sensor for sensing analyte, a method and system thereof – Indian Patent.
- A method for obtaining a coating of a metal compound onto a substrate, an apparatus and a substrate there of – Indian Patent.
- Compliant platforms to generate amplified displacements, compliant platform for sensing applied motion and method of designing DaCM – Indian Patent.
- Modification of Fermi-level pinning behaviour at the Germanium surface through sulfur passivation treatment – Indian Patent.

- A technique to prepare brittle nano crystals of noble metals – Indian Patent.
- Idea sustain – It is a computer aid based on a potential impact estimation rationale capture technology to help designers develop environmental friendly product life cycle. Indian Patent.
- Idea Inspire – It is a computational tool for supporting designers to generate novel solutions for product designing problems by providing as stimuli information about natural or artificial systems that are analogically relevant. Indian Patent.
- A composition of electrode material in the form of a coating and a process thereof – Indian Patent.
- Room temperature synthesis of coinage nanoparticles – Indian Patent.

4.11 NANOTECHNOLOGY COURSES

Few Colleges and in India:

- B.Tech & B.Tech + M.Tech in Nanotechnology (Dual Degree) – Amity Institute of Nanotechnology, Noida www.amity.edu/ http://www.amity.edu/aint/
- B. Tech in Nanotechnology – Shri Guru Granth Sahib World University, Punjab
  http://sggswu.org/department/detail/dept-of-nanotechnology
- B. Tech Nanotechnology – SRM University, Chennai
  http://www.srmuniv.ac.in/engineering/school-of-basic-sciences/department-physics-and-nanotechnology/about-the-department
- M.Sc. Nanoscience & Technology – Allagappa University, Tamilnadu
  http://alagappauniversity.ac.in/academic/coursesoffered.php
- M. Tech Nanotechnology – SRM University, Chennai
  http://www.srmuniv.ac.in/engineering/school-of-basic-sciences/department-physics-and-nanotechnology/about-the-department
- Amity Institute of Nanotechnology, Noida, www.amity.edu/aint
- M.Sc. – Nanoscience by Research + M.Tech – Nanotechnology (Dual Degree) – Amity Institute of Nanotechnology, Noida
- M.Sc. – Forensic Nanotech &Nanobiotechnology

- M.Tech – Nanotechnology
- Amrita Vishwa Vidyapeetham University, Coimbatore
- M. Tech in Nanotechnology & Renewable Energy
- M. Tech in Naomedical Sciences
  https://www.amrita.edu/academics

- Jawarharlal Nehru Technological University, Hyderabad
- M.Tech Nanotechnology
- M.Sc Nano Science and Technology
  http://jntuh.ac.in/new/academic/nano-science.html

- Sastra University, Thanjavur, TM
- M. Tech in Medical Nanotechnology
- M. Tech in Nanoelectronics
  http://www.sastra.edu/index.php/2014-01-29-07-16-11/postgraduate

- Shri Guru Granth Sahib World University, Punjab
- M. Tech in Nanotechnology
- M. Tech in Nanotechnology (5 year Integrated)
  http://sggswu.org/department/detail/dept-of-nanotechnology

- University of Rajasthan, Jaipur
- Masters of Nanotechnology (5 Years Integrated)
- Masters of Nanotechnology (3 Years)
- Masters of Nanotechnology (2 Years)
  http://www.uniraj.ac.in/cct/index.php?pid=8
- M. Tech in Nanotechnology – VIT University, Vellore, TN
  http://www.vit.ac.in/admissions/coursepage/1/22
- M. Tech Nanoscience and Nanotechnology – Delhi University, New Delhi
  http://admission.du.ac.in/pgadm/index.php?page=nano-science

Ph.D. Degree Programs
- Ph.D in Nanoscience and Technology – Alagappa University, Tamilnadu
  http://alagappauniversity.ac.in/departments/programs.php?dept_id=29
- Ph.D. in Nanoscience and Nanotechnology (Full Time and Part Time) – Amity Institute of Nanotechnology, Noida
  http://www.amity.edu/aint/

- Ph.D. in Nanomedical Sciences – Amrita Vishwa Vidyapeetham University, Coimbatore
  https://www.amrita.edu/academics
- Ph.D. in Nanoscience & Technology – Aryabhatta Knowledge University, Patna
  http://akubihar.org/acnn/index.html
- Ph.D. in Nanotechnology – Shri Guru Granth Sahib World University, Punjab
  http://sggswu.org/department/detail/dept-of-nanotechnology
- Ph.D. in Nanotechnology – University of Rajasthan, Jaipur
  http://www.uniraj.ac.in/cct/index.php?pid=8

Few Colleges and Graduate Programmes – Abroad
- B.S. Nano Engineering – US San Diego, Jacob School of Engineering, CA
  http://nanoengineering.ucsd.edu/undergrad-programs/degree
- B.S. Chemistry with a concentration in Nanotechnology – Carleton University, Canada
  http://carleton.ca/chemistry/prospective-students/undergraduate/degree-and-programs/
- B.Sc Nanoscience – Carleton University, Canada
  https://admissions.carleton.ca/programs/nanoscience/
- B.Sc Materials Engineering with specialization Nanotechnology – Drexel University (USA)
  http://www.materials.drexel.edu/academics/undergrad/bs/tracks/
- B.Sc Program – DTU Nanotech, Denmark
  http://projects.nanotech.dtu.dk/projects.asp?projecttype=bachelor#P1606
- B.Sc (Hon) Nanotechnology – Flinders University, Australia
  http://www.flinders.edu.au/courses/undergrad/bscnnh/bscnnh_home.cfm
- B.Sc Nanoscience – Heriot Watt University, UK
  http://www.undergraduate.hw.ac.uk/programmes/CF10/#introduction
- B.Sc in Nanosystems Engineering – Louisiana Tech University, USA
  http://coes.latech.edu/nanosystems-engineering/bs-nanosystems-engineering.php

- B.Sc in Nanotechnology – Leibniz University Hannover, Germany
  http://www.lnqe.uni-hannover.de/study_nano_bachelor.html?&L=1
- Bachelor of Nanotechnology/ Bachelor of Science – La Trobe University, Australia
  http://www.latrobe.edu.au/courses/nanotechnology
- B.Sc Microtechnology and Nanostructures – Saarland University, Germany
  http://www.uni-saarland.de/en/campus/study/academic-programmes/first-degree-programmes/microtechnology-and-nanostructures-bachelor.html
- B. Sc Science (Nanotechnology) Program – RMIT University, Australia
  http://www.rmit.edu.au/study-with-us/levels-of-study/undergraduate-study/bachelor-degrees/bp247/#pageId=overview
- Bachelor Program Nanoscience and Nanotechnology – Universidade Federal do Rio de Janeiro – Brazil
  http://www.nano.ufrj.br/
- Bachelor in Nanoscience – University Hamburg, Germany
  http://www.nano.uni-hamburg.de/index.html
- Bachelor in Science Nanotechnology– University Erlangen-Nurnberg, Germany
  http://www.nano.studium.uni-erlangen.de/index.shtml
- B.Sc in Nanoscience – University of Basel, Switzerland
  http://www.nccr-nano.org/nccr/study/
- B.Sc in Nanoscience and Nanotechnology – University of Copenhagen, Denmark
  http://nano.ku.dk/english/study/Bachelor/
- Master of Nanotechnology/ Bachelor of Science – La Trobe University, Australia
  http://www.latrobe.edu.au/courses/nanotechnology/postgraduate
- Master's Programme in Micro and Nanotechnology – Aalto University, Finland
  http://www.micronova.fi/education/master-s_programme_in_micro_and_nanotechnology/
- Professional Science Master (PSM) Program in Nanoscience – Arizona State University, USA
  http://nanoscience.asu.edu/

- M.Sc Nanotechnology and Microfabrication – Bangor University, UK
  http://www.bangor.ac.uk/eng/postgraduate-courses/nanotechnology-and-microfabrication-msc
- Master Programme in Micro and Nano Systems – Chemnitz University of Technology, Germany
  https://www.tu-chemnitz.de/etit/studium/stugang/english-master-mns.php
- M.Sc Microsystems and Nanotechnology – Cranfield University, UK
  http://www.cranfield.ac.uk/courses/masters/applied-nanotechnology.html
- Master of Science Nanoscience – Delft University of Technology, Netherlands
  http://www.tudelft.nl/en/study/master-of-science/master-programmes/nanoscience/
- M.Sc Program – DTU Nanotech, Denmark
  http://projects.nanotech.dtu.dk/projects.asp?projecttype=master
- Master of Nanotechnology – Flinders University, Australia
  http://www.flinders.edu.au/courses/postgrad/nt/nt_home.cfm
- Master of Nanoscience and Nanotechnology – Erasmus Mundus, Belgium
  http://www.emm-nano.org/
- M. Phil. Nanoscience and Technology – Hong Kong University of Science and Technology, China
  http://nanoprogram.ust.hk/intro.html
- MRes in Nanomaterials – Imperial College of London, UK
  http://www.imperial.ac.uk/study/pg/courses/chemistry/nanomaterials/
- M.Sc Nanomolecular Science – Jacob University of Bremen
  http://www.jacobs-university.de/ses/nanomol
- M.Sc Program Nanotechnology – Istanbul Technical University
  http://www.nano.itu.edu.tr/Content.aspx?code=education
- Master of Science in Nanoengineering – Joint School of Nanoscience and Nanoengineering, North Carolina, USA
  http://jsnn.ncat.uncg.edu/academic/nanoengineering/master-of-science-in-nanoengineering/
- Master's Programme in Nanotechnology – KTH – Royal Institute of Technology, Sweden
  https://www.kth.se/en/studies/master/kth/nanotechnology

- Master of Nanotechnology – La Trobe University, Australia
  http://www.latrobe.edu.au/courses/nanotechnology/postgraduate
- M.S. in Molecular Sciences and Nanotechnology – Louisiana Tech University, USA
  http://coes.latech.edu/grad-programs/msnt.php
- International Master Program "Engineering Nanoscience" – Lund University, Sweden
  http://www.lth.se/english/education/master/nanoscience/
- Nanotechnology Master Program – Nile University
  http://www.nileu.edu.eg/nano/education.html
- Master of Nanoscale Engineering – Lyon Institute of Nanotechnology, France
  http://master-nano.universite-lyon.fr/
- Master of Science (Nanoscience) – Massey University, New Zealand
  http://www.massey.ac.nz/massey/learning/programme-course-paper/programme.cfm?prog_id=92431&major_code=2796

Ph.D. Degree Programs
- Ph.D. in Molecular Sciences and Nanotechnology – Louisiana Tech University, USA
  http://coes.latech.edu/grad-programs/msnt.php
- Ph.D. in Micro and Nano Technology – Lancaster University, UK
  http://www.lancaster.ac.uk/engineering/postgraduate/research-courses/
- Ph.D. in Nanoengineering – Joint School of Nanoscience and Nanoengineering, North Carolina, USA
  http://jsnn.ncat.uncg.edu/academic/nanoengineering/ph-d-in-nanoengineering/
- Ph.D. Micro and Nanotechnology – Middle East Technical University, Turkey
  http://mnt.metu.edu.tr/
- Ph.D. Program NanoCore – National University of Singapore, Singapore
  http://www.nanocore.nus.edu.sg/positions.html
- Ph.D. in Nanoscience – Murdoch University, Australia
  http://www.see.murdoch.edu.au/areas/nanoscience/postgrad/higher.html

- Ph.D. Nanoengineering – University of California, San Diego, USA
  http://nanoengineering.ucsd.edu/graduate-programs/degree/nanoengineering/doctoral
- Ph.D. Nanoscale Science – University of Albany, New York, USA
  http://www.sunycnse.com/PioneeringAcademics/GraduatePrograms/NanoscaleScienceProgram.aspx
- Ph.D. School in Nanostructures and Nanotechnologies – University of Milano Bicocca, Italy
  http://www.nano.unimib.it/

4.12 NANOTECHNOLOGY IN SCHOOLS

Nanoscience and nanotechnologies offer a new instrument to bring exciting science and technology into the classroom. Nanotechnologies are now used in numerous devices with which young students are very familiar, such as computers, mobile phones and iPads. Nanoscience offers the possibility to improve numerous material properties and create new ones. In the future, more and more products will incorporate some form of 'nano' either a nanomaterial, or a nano-enabled technology. Bringing 'nano' into the classroom means bringing in the latest cutting-edge science and technology and talking about very exciting future scientific developments. Number of schools in US and in other countries are introducing nanoscience and nanotechnology in schools. In the United States, few high schools offer a two-semester course in nanotechnology. In addition, "nano" concepts are taught during traditional science classes using a number of educational resources and hands-on activities developed by dedicated non-profit organizations, such as

- The National Science Teacher Association, which has published a number of textbooks for nanotechnology in K-12 education, including a teacher's guide and an activity manual for hands-on experiences.
- Nano-Link, a notable program of the Dakota County Technical College, which has developed a variety of nanotech-related hands-on activities supported by toolkits to teach concepts in nanotechnology throughout direct lab experience.
- Omni Nano, which is developing comprehensive educational resources specifically designed to support a two-semester high school course,

both online and in classrooms. Omni Nano also discusses issues in nanotechnology education on its dedicated blog.
- Nano4Me, which has a good amount of resources for K-12 education. Their K-12 resources include introductory level modules and activities, interactive multimedia, and a collection of experiments and hands-on activities.
- Nanoscale Informal Science Education Network (NISE), which has a website of educational products designed to engage the public in nano science, engineering, and technology.
- in2nano, is a high school outreach program in Egypt, aiming to increase scientific literacy and prepare students for the sweeping changes of nanotechnology.

4.13 RISK OF NANOTECHNOLOGY

Broad range of applications and varieties of industrial sectors affected by nanotechnologies as discussed above, it is evident, that the risk landscape associated with nanotechnologies will be complex rather than a homogenous set of risks. To name just a few:
- Risks related to the protection of intellectual property,
- Risks regarding the impact on the economic development of countries
- Risks regarding privacy when miniature sensors become ubiquitous
- Environmental risks from the release of nanoparticles into the environment
- Safety risks from nanoparticles for workers and consumers
- Futuristic risks like human enhancement and self replications of nano machines.

Almost all safety concerns that have been raised about nanotechnologies are related to free rather than fixed engineered nanoparticles. The risk and safety discussion related to free nanoparticles will concern only a fraction of the applications of nanotechnologies. In most applications nanoparticles will be embedded in the final product and therefore not come into direct contact with workers, consumers or the environment. They are unlikely to raise concerns because of their immobilisation. Exceptions are possible when the products or materials within which nanoparticles are enclosed are discarded, burned or otherwise destroyed.

Positive Effects: A fair assessment of the risks of any new technology must also consider positive contributions to human health and environmental safety in many ways. Nanotechnology has the potential to contribute to solve urgent issues like the provision of clean drinking water or more efficient energy conversion and energy storage beside making high contribution to economic benefits, the potential to create jobs, wealth and wellbeing. In the area of medicine, there are quite a few technological developments that promise enhanced diagnostic possibilities, new ways to monitor patients, new ways to treat diseases like cancer and to reduce side effects. To give a few examples:

- Nanoparticles can be used as carriers for targeted drug delivery. Their ability to penetrate certain protective membranes in the body, such as the blood- brain barrier, can be beneficial for many drugs. This could open the way for new drugs from active substances that have not been able to pass clinical trials due to less precise delivery mechanisms,
- Nanosensors and lab-on-a-chip-technologies will foster early recognition and identification of diseases and can be used for continuous monitoring of patients with chronic diseases,
- New therapeutic methods for the treatment of cancer with the help of nanoparticles are investigated.
- Ultrasensitive detection of substances will improve safety parameters in many areas such as industrial medicine, environmental medicine and food safety. Bacterial pathogens can be detected in very low concentrations with the help of nanoparticles. New findings indicate that specially treated nanoparticles could allow to detect a single E. coli bacterium in a meat sample.
- The potential benefits for our environment range from resource efficient technologies reducing waste, to new ways to transform and detoxify a wide variety of environmental contaminants, such as chlorinated organic solvents, organochlorine pesticides, and PCBs.

Negative Effects: When bulk materials are made into nanoparticles, they tend to become chemically more reactive. Even chemically inert materials like gold or platinum are able to catalyse chemical reactions in nanopowder form. Nanoparticles generally are more toxic when incorporated into the human body than larger particles of the same materials. The biggest concern is that free nanoparticles or nanotubes could be inhaled, absorbed through the skin or ingested. Inhaled particles can have two major effects on the human body. Primary toxic effect is to induce inflammation in the respiratory tract, causing tissue damage and subsequent systemic effects. Transport through the blood stream to other vital organs or tissues of the body may result in cardiovascular and other extra-pulmonary effects.

From animal experiments and analogies to studies on incidentally produced ultra fine particles such as from burning of fuels, it is possible that at least some nanoparticles are hazardous for the human body and that the exposure to these nanoparticles should be avoided or at least minimised. One promising path to prevent potential health hazards proposed by a number of scientists is to make the particles biodegradable. Particles that are degradable either by water or by lysis with enzymes will greatly reduce the risks involved because they do not persist in the body. It is evident that the slower the particles are cleared, the higher the tissue burden can be. With a short bio durability, long-term effects can be minimised or even excluded. Biocompatability can serve as one major engineering parameter for nanomaterials in the future.

To summarise, the information about nanoparticles and the environment is at an early stage of research like the effect of nanoparticles on humans, other than humans, how they behave in the air, water or soil, or their ability to accumulate in food chains. Taking into account the varieties and parameters like size, shape, specific surface treatment, chemical composition of nanoparticles, considerable research are still needed to close the knowledge gaps.

4.14 ETHICAL, LEGAL AND SOCIETAL ISSUES

Responsible development of nanotechnology is one of the goals of the National Nanotechnology Initiative (NNI). An important component of responsible development is the consideration of the ethical, legal, and societal implications. One potential problem with nanotechnologies is how it is being implemented and commercialized. While with introduction of new technology like nuclear energy, the subject is very deliberate and hotly debated, nanotechnologies are creeping into our lives more or less unnoticed and to a large degree unquestioned. Many people haven't heard about it or don't care enough to get informed about it. Industry is picking the most promising research results for commercialization in its effort to improve its products. The electronics industry, being well on its way to become a nanotechnology industry, is an excellent example of how this "nanotechnology creep" is happening,  unquestioned and well accepted by consumers. Ever increasing appearance of internal and external enhancements of the human body, driven by lucrative and big business sport on one hand and the medical technology to deal with injuries on the other is enabling a culture of increasing demand for, and acceptance of, improvements to and modifications of the human body (structure, function, abilities) beyond its species-typical boundaries. The question of whether medical and electronic nanotechnologies should be used to make intentional changes in or to the body when the change is not medically necessary is just one hot topic in a long list of concerns. The good news is that these questions are being asked. NNI is committed to fostering the development of a community of experts on ethical, legal, and societal issues (ELSI) related to nanotechnology and to building collaborations among communities, such as consumers, engineers, ethicists, manufacturers, nongovernmental organizations, regulators, and scientists. These stakeholders will consider potential benefits and risks of research breakthroughs and provide their perspectives on new research direction.

4.15 CONCLUSION

Kurzweill had visioned that tiny robot "assemblers" will swarm unseen in the air, water, and even our bodies, building anything from the atoms up. It's still mostly a dream. But if we look at "life", these nanoassemblers are already here. Basic organic machinery is ceaselessly pulling in raw materials and building, molecule by molecule, cells, organs, and bodies—from hearts to brains, ants to blue whales. Each of us is alive today because of countless nanobots operating within each of our trillions of cells. We give them biological names like a "ribosome," but they are essentially machines programmed with a function like "read messenger RNA to create a specific protein." That being said, it's important to distinguish between "wet" or "biological" nanotech which basically uses DNA and the machinery of life to create unique structures made of proteins or DNA (as a building material) and a more Drexlerian (K. Eric Drexler, 1986, Engines of Creation: The Coming Era of Nanotechnology) nanotech which involves building an "assembler," or a machine that can 3D print with atoms at a nanoscale and effectively create any structure that is thermodynamically stable. We draw analogies between our inventions and the natural world. These analogies help us visualize and develop systems. What life does on the smallest scales and what nanotechnology aims to do aren't very different. Both assemble and repair microscopic structures by taking in atoms and molecules in one state and transmuting them into a different one. Life does its own thing, tuning and retuning itself to suit the environment over eons. Nanotechnology, on the other hand, would be tailored to our purposes by designing microscopic machines programmed at our will. We don't yet have sleek nanoassemblers, but biology will be a bridge to that future. Research at the convergence of biotechnology and nanotechnology is taking control of life's "nanoassemblers" and repurposing them.

This multidisciplinary field of nanotechnology is bringing the science of the incomprehensibly small devices closer to reality. The effects of these developments will be so vast that they will probably affect virtually all fields of science and technology holding the promise of delivering the greatest technological breakthroughs in history. On the flip side the potential of nanotechnology in chemical and biological weapons is equally disquieting as nanotechnology can considerably enhance the delivery mechanism and toxic substances. The ability of nanoparticle to penetrate the human body and its cells could make biological and chemical warfare much more potent, and easier to manage. So, nanotechnology besides being a boon does have the potential to become a bane of the society. Precautionary Principle is often invoked when dealing with situations that might be hazardous. The Center for Responsible Nanotechnology (CRN) recognizes the possibility of grey areas, along with the possibility of dangerous arms races and widespread use of destructive products. CRN is dedicated to reducing these risks as far as possible.

REFERENCES

1. Essentials of Nanotechnology, Jeremy Ramsden, © 2009 Jeremy Ramsden &Ventus Publishing ApS ISBN 978-87-7681-418-2
2. Nanofrontiers, Vision for the Future of nanotechnology, Karen F Schimdt
3. Nanomanufacturing- Perspective and Applications, CIRP Annals 2017, Fang, F. Z.; Zhang, X. D.; Gao, W.W.; Guo, Y. B.; Byrne, G.; Hansen, Hans Nørgaard
4. Opportunities and Risk of Nanotechnology, Report in OECD International Future Programme
5. Nanorobots: Where We Are Today and Why Their Future Has Amazing Potential, By Peter DiamandisPeter, May 16, 2016
6. How Nanotech Will Lead to a Better Future for Us All, By, Alison E Berman, Aug 12, 2016
7. How Scientists Are Hacking Biology to Build at the Molecular Scale, By Jason Dorrier, May 10, 2016
8. https://en.wikipedia.org/wiki/Nanotechnology
9. https://en.wikipedia.org/wiki/Nanomaterials
10. Singularity Hub www.singularityhub.com

CHAPTER 5
GENETICS

INTRODUCTION

On April 25, 2003, a group of geneticists representing six countries announced that it had mapped every one of the three billion letters making up the human genome. What had taken scientists of the Human Genome Project 13 years and $3 billion to achieve can be done today for $3,000. With thousands of human genomes now sequenced, one of the principal leaders of the Human Genome Project, Eric S. Lander, says that in just five years the world could have "a complete catalogue for most of the important diseases." In an interview with Divya Gandhi (divya.gandhi@thehindu.co.in), Professor Lander, President and Founding Director of the Board Institute of the Massachusetts Institute of Technology and Harvard University, has talked of potential of genetics study in India. "Indian science allocation has not grown in comparison to China and allocation in science and technology has to increase so that the country can take a lead in this area. But it is important to say that it is not all about money. You need great scientists; you need to give them their autonomy. Science should be directed by scientists and scientific priorities. It has never really worked for the political branch to pick and choose projects. This country trains extraordinary scientists and we in the U.S. get the benefit. Your brain drain is our brain gain. India is certainly active. But is India doing as much as it should be doing? No. India is perhaps the single most interesting country to study genetics. There is tremendous diversity across India. You have a country of over a billion people with so many different ethnic groups; endogamous groups with people tending to marry within particular groups; and parts of India with consanguinity marriages with relatives. You have amazing genetic variation — more than any other country perhaps in the world. And there is so much that can be learnt from that. The excellent science that is going on is only a fraction of what could go on. India should be one of the models in studying genetic variation and how it relates to disease. So, I would be in favour of seeing much greater activity in genomic medicine in India."

Genetics and its allied disciplines are set of the technologies where pace of change is exponential, fast and far-reaching and could irreversibly alter human existence. Human brain power, knowledge, skills, and personality quirks are getting combined with computer power in order to think, reason, and communicate. This merger of man and machine, coupled with the sudden explosion in machine intelligence and rapid innovation in gene research and nanotechnology, is resulting in a world where the distinction between the biological and the mechanical, or between physical and virtual reality is fading fast. Technological revolutions will allow us to transcend our frail bodies with all their limitations. Illness, as we know, will be eradicated. Human existence will undergo a quantum leap in evolution. By understanding the information processes underlying life, we will learn to reprogram our biology to achieve the virtual elimination of disease, dramatic expansion of human potential, and radical life extension. Genetic and molecular science will extend biology and correct it's obvious flaws like vulnerability to disease. We are rapidly gaining the knowledge and the tools to drastically extend the usability of the house each of us calls our body and brain. Nanomedicine researcher Robert Freitas (https://en.wikipedia.org/wiki/Robert_Freitas) estimates that eliminating 50% of medically preventable conditions would mean extending human life expectancy to 150 years and by preventing 99% of naturally occurring medical problems, life expectancy could lead to increase to 1,000 years. Drug discovery was once a matter of finding chemicals that produced some beneficial result without excessive side effects. Today drugs are designed to carry out precise missions at the molecular level. Gene technologies will control how genes express themselves. Many new therapies currently in development and testing are based on manipulating peptides either to turn off the expression of disease-causing genes or to turn on desirable genes that may otherwise not be expressed in a particular type of cell. Accelerating progress in biotechnology will enable us to reprogram our genes and metabolic processes to propel the fields of genomics, proteomics, gene therapy, rational drug design, as well as the therapeutic cloning of rejuvenated cells, tissues, and organs. Before we proceed with the details of this amazing technology, let us see some examples.

A couple sits close, intently studying a dossier which has a list starting with Embryo #1. According to the description, Embryo #1 is female, has a high risk of Type II Diabetes, will have blue eyes and blond hair, and has a 20% chance of being in the 90th percentile for math ability. Alternatively, Embryo #100 is male, will have blue eyes and dark hair, has a 60% chance of being in the top 10% for musical ability, and is at a high risk for depression. Between Embryo #1 and #100 lie similarly detailed descriptions. While this may sound like science fiction, according to Hank Greely, Dean F and Kate Edelman, Johnson Professor of Law at Stanford University, this scenario is the soon-to-be future of human reproduction.

"I predict in [my] book *The End of Sex and the Future of Human Reproduction,* that in 20 to 40 years the majority of babies born to people with good health coverage anywhere in the world will not be conceived in a bed or in the back of a car or under a 'keep off grass' sign, but will be conceived in a lab so that parents can then do whole genome sequencing and pick the embryo that they want," Prof. Greely said during his Gladstone GO Graduate Student Organization sponsored bioethics seminar on April 14, 2017.

A family sat on the top of the world and gazed at the stars. They lived on the Tibetan Plateau, 4200 m (14,100 ft) above sea level, in a site now known as Chusang. Although far from the comfort of climate of sea level, a hot spring provided a welcoming buffer against the chilled air. At night, the family lit fires in a hollow built into the slope. Living at the centre of the plateau, they simply couldn't migrate up and down the mountain with the seasons as other Tibetan people did during this period. They were here year-round, enduring the heavy snowfall, biting winds, and encroaching glaciers of winter. Their survival is extraordinary. While the heat of fire could protect them from the cold, the family at Chusang couldn't shelter from an obvious yet insurmountable obstacle of living on the plateau: the air becomes thinner with every step towards the sky. At more than 4,000 m (13,000 ft) above sea level, each breath contains around a third less oxygen than the same breath far below. But deep inside each of their bodies,

within their blood and DNA, an ancient and unique trick to surviving at altitude protected them from the thin air in which they built their home. Any mountain climber will be able to describe the shortness of breath that normally comes with altitude. It's not that the air has a lower percentage of oxygen it's around 21% wherever you stand in the world. But air pressure decreases the further you walk or fly from the sea's surface, allowing the gas molecules to spread out in all directions, and a lung can only stretch so far to compensate. There are ways to deal with this change in pressure, however. Over many hundreds of generations, people living on the Andean altiplano that extends from Peru into Bolivia, have evolved barrel-shaped chests that increase the volume of each of their breaths. And since the late 1800s, scientists have known that their blood is pumped full of red blood cells and haemoglobin, the oxygen-carrying molecules. When the air is thin, the blood thickens to increase the amount of oxygen it can shepherd to cells around the body. This hematopoietic (literally, "blood" and "to make" in Greek) response is also found in anyone who decides to hike up a mountain. It was only in the late 1970s and early 80s, after hiking to seven villages in Nepal, that Cynthia Beall, an anthropologist from Case Western Reserve University in Ohio found that Tibetans didn't fit this theory. Firstly, they lacked the barrel-shaped chests, but seemed to breath at a faster rate than Andeans. And second, in the autumn of 1981 Beall and her colleagues found that Tibetans have surprisingly low haemoglobin levels, often within the range of what is normal for people who live at sea level. Although they live on the so-called "roof of the world", their physiological state seemed surprisingly similar to those who had never left its floor. "At first, this was anxiety inducing," says Beall. "You think, 'Oh gosh, did I measure the wrong people? Did I do the wrong measurements? Is there something I'm missing?" But after returning to Tibet and Nepal many times since, collecting more data from more villages, she only found support for her initial results: at high altitude, low-oxygen environments, Tibetan people reduce the amount of oxygen their blood can carry. How could this be? What at first appears to be highly paradoxical – not to mention potentially dangerous – actually makes a lot of sense, protecting Tibetan people from some of the nastier side effects of the high-life.

One benefit, for instance, is reduced wear and tear on their blood vessels. "If you have high levels of haemoglobin your blood tends to be more viscous, and that can have a lot of damaging effects," says Tatum Simonson from the University of California in San Diego. "You're basically pumping this very thick, concentrated blood throughout your system. Your heart is on overdrive." A possible outcome of this added stress on the entire circulatory system is chronic mountain sickness, or CMS. As with short-term altitude sickness, the remedy for CMS is a slow descent into thicker, more oxygenated air. But it is no cure. Fluid may have already built up in the lungs (a high altitude pulmonary oedema, or HAPE) or in the brain (a high altitude cerebral oedema, or HACE), or the thick blood may be congested in other vital organs. The worst-case scenario is death. Certainly thin blood helps reduce CMS risk, but it's certainly not the only reason Tibetan people can live happily at such extremes. In 2005, for instance, Beall and her colleagues found that Tibetans exhale more nitric oxide compared to people living in at sea level. Originally described as a relaxation factor, this gas leads to a widening of blood vessels in the lung and around the body, known as vasodilation. With more space, blood flow and oxygen transport can increase. And, as Simonson suggests, what if Tibetans simply don't require as much oxygen as other people? What if their muscles are just more efficient with their usage, for instance? "Perhaps they are already so well tuned that they don't need more red blood cells and haemoglobin to bind more oxygen," she says. Her work is now exploring this possibility. Although she has visited the Tibetan Plateau several times for her research, Simonson surveys the history of this region back in her laboratory. As a geneticist, she can scour the genomes (the entire DNA sequence of an individual) of Tibetan people to find what underlies their unique adaptations to the high life. In 2010, by comparing the genomes of 30 Tibetan people to those from a Han Chinese population living in Beijing, Simonson could identify those genes that were associated with living at high-altitude. Simonson's lab wasn't the only one attempting this. In the space of two weeks in 2010, a total of three research groups each published a study that found a handful of genes that were markedly different between the two populations. Of note, two genes called EPAS1 and EGLN1 stood out from the crowd, and, importantly, were already known to modulate the haemoglobin levels in blood. After looking more closely at the EPAS1 gene from the Tibetan genomes, Nielsen not only found it was a steep change, but it was a unique one too. After searching through the aptly named 1,000 Genomes Project, he couldn't find anything quite like it elsewhere. "The DNA sequence that we saw in Tibetans was simply too different," Nielsen says. It was as if Tibetans had inherited the gene from another species. And, in fact, that's exactly what had happened. For people of Asian ancestry Nielsen instead looked to Denisovans, another branch of the human family tree. Discovered in the Altai Mountains in Siberia, they are known only from two teeth, a tiny finger bone, from which Paabo and his colleagues published a rough genome in 2012. The results demonstrated that populations from Papua New Guinea, Australia, and a few regions of southeast Asia had inherited between 1-6% of their genomes from Denisovans. It was a case of third time lucky. "There was a complete match," he says. "It's so hard to believe that it could possibly be true. But it is." Between 50,000 and 30,000 years ago, some Denisovans and the ancient ancestors of Tibetan and Han Chinese people had sex, merged their genomes, shuffled the genes like a deck of cards, and produced children who would grow up to have offspring of their own. Over the next tens of thousands of years, this gene seems to have conferred little benefit to Han Chinese people and is only found in roughly 1% of the population today. But for all those intrepid groups that moved up onto the Tibetan Plateau, including the Chusang family, it helped make every breath easier, every heartbeat less dangerous. On the Tibetan Plateau, 78% of the population has this version of EPAS1, a gene that separates them from those far below, but connects them to the past. Over 50,000 years in the making, this story still doesn't have an ending. Although its origin is known, those areas that make EPAS1 in Tibetans unique are still largely unchartered. The specific change (or changes) that leads to a reduction in haemoglobin content is still unknown. "All the geneticists say it's in an area that's very hard to sequence," says Beall. The new explorers, those of mountains of data and genomes, still have a long journey ahead of them. (Alex Riley, a writer based in Berlin, Germany. He tweets as @riley__alex).

5.1 DEFINITIONS: AT THE VERY OUTSET, LET'S DEFINE FEW TERMS

What is a Gene: In brief, it is a molecule of DNA, present in our every cell. But the richness of the gene concept goes much beyond this simple definition. We have come to feel that they are fundamental to us. But there is also uncertainty and confusion about what this means. For example, does having cancer in a first degree relative mean I have 'cancer genes'? If I have genes also found in Stone Age fossils, does this mean I am very primitive?

Genetic and Genetic Engineering: Genetics are the building blocks of life. They're the code for every living organism on Earth. Genetic engineering is manipulation of that code which recombines the DNA of two species. Genetics is a science and genetic engineering is application of the science to living beings. It is the deliberate, controlled manipulation of the genes in an organism with the intent of making that organism better in some way. This is usually done independent of the natural reproductive process and is done by a set of technologies to change the genetic makeup of cells, including the transfer of genes within and across species boundaries to produce improved or novel organisms.

Genetic Engineering Tools: To do genetic engineering, scientists use tools of DNA technology. Scientists have modified bacteria to produce insulin for diabetics, have modified corn to be resistant to herbicides for less harmful farming and have modified mice to grow human cancer tumours to test medications. The most common method of genetic engineering is to snip out a piece of DNA from one organism and replace it with a section from another organism. That's called recombinant DNA, and it's done with the help of a couple of different molecules used to cut apart and glue together DNA molecules.

GMO: The product of genetic engineering is a Genetically Modified Organism (GMO). Any change in the genome is considered natural if it is a result of mutation or selective breeding. The genetic modification may be either through human manipulation or as a result of traditional plant breeding methods like selective breeding or crossbreeding between plants within the same species.

Cloning: Whereas cloning produces genetically exact copies of organisms, genetic engineering refers to processes in which scientists manipulate genes to create purposefully different versions of organisms and, in some cases, entirely new living things.

Genetic Engineering vs DNA Technology: There's a very subtle difference between DNA technology and genetic engineering. Genetic engineering refers to techniques which are used to modify the genotype of an organism to change its phenotype. That is, genetic engineering manipulates an organism's genes to make it look or act differently. DNA technology refers to the methods used to modify, measure, manipulate and manufacture within the DNA molecule. Because genes are stored in DNA, genetic engineering is done with DNA technology. But DNA technology can be used for more than genetic engineering.

Biomedical/Bioengineering: It is application of traditional engineering techniques in subspecialties such as mechanical, materials, or electrical to solve biological or clinical problems. So the difference is as clear as, say between chemistry and chemical engineering, where one is theory and the other is practice.

Synthetic Biology Vs. Genetic Engineering: Synthetic biology, one of the newest areas of research in biology, aims to engineer new biological systems in order to create organisms with novel human-valuable features. This term has many times been called "extreme genetic engineering". The question arises if this problem is a matter of terminology or if there is a scientific difference between these two fields of biology. Even though DNA manipulation is the basis of these two practices, it is important to clarify that genetic engineering is the set of methodologies for altering the genetic material, while synthetic biology is a new field that uses genetic engineering, mathematical modelling, and industrial analysis among other tools in order to create new biological systems. Genetic engineering is one of the multiple fields essential for synthetic biology which has direct social purposes, namely, to develop a novel therapy for treating cancer and a novel process to produce clean energy from air pollution, respectively. As a consequence many other fields have to be incorporated to synthetic biology. So genetic engineering is a useful tool for scientists, and is fundamental for synthetic biology.

Genetic Engineering vs Biotechnology: Genetic engineering deals entirely with study of genes and genomics. This field broadly deals with deciphering the puzzles of gene by manipulation of genetic material. Genes of the organism can be modified so as to persuade the organism to do any particular function according to our need. Biotechnology, on the other hand, is a very broad field which uses genetic engineering in order to accomplish different aims for the benefit of beings. Though a very large part of biotechnology depends upon genomics it includes many other branches also.

5.2 GENETICS, GENE AND DNA

We study genetics for two basic reasons. First, genetics occupies a pivotal position in the entire subject of biology. Therefore, for any serious student of plant, animal, or microbial life, an understanding of genetics is essential. Second, genetics is central to numerous aspects of human affairs. It touches humanity in many different ways. Indeed, genetic issues seem to surface daily in our lives, and no thinking person can afford to be ignorant of its discoveries. Some define it as the "study of heredity," but hereditary phenomena were of interest to humans long before biology or genetics existed as the scientific disciplines that we know today. Ancient peoples were improving plant crops and domesticated animals by selecting desirable individuals for breeding. They also must have puzzled about the in- heritance of individuality in humans and asked such questions as "Why do children resemble their parents?" and "How can various diseases run in families?" But these people could not be called "geneticists." Genetics as a set of principles and analytical procedures did not begin until the 1860s, when an Augustinian monk named Gregor Mendel performed a set of experiments that pointed to the existence of biological elements that we now call genes. The word genetics comes from the word "gene," and genes are the focus of the subject. Whether geneticists study at the molecular, cellular, organismal, family, population, or evolutionary level, genes are always central in their studies. Simply stated, genetics is the study of genes.

Gene: The words, Gene, Genetics, Genome, and Genomics, are all derived from a Greek word, gen, meaning birth or origin. A gene is a section of a threadlike double-helical molecule called deoxyribonucleic acid, abbreviated DNA. The discovery of genes and the understanding of their molecular structure and function have been sources of profound insight into two of the biggest mysteries of biology:

> What makes a species what it is? We know that cats always have kittens and people always have babies. This common sense observation naturally leads to questions about the determination of the properties of a species. The determination must be hereditary because, for example, the ability to have kittens is inherited by every generation of cats.

> What causes variation within a species? We can distinguish one another as well as our own pet cat from other cats. Such differences within a species require explanation. Some of these distinguishing features are clearly familial; for example, animals of a certain unique colour often have offspring with the same colour, and in human families, certain features, such as the shape of the nose, definitely "run in the family." Hence we might suspect that a hereditary component explains at least some of the variation within a species.

The answer to the first question is that genes dictate the inherent properties of a species. The products of most genes are specific proteins which are the main macromolecules of an organism. When we look at an organism, what is seen is either a protein or something that has been made by a protein. The amino acid sequence of a protein is encoded in a gene. The timing and rate of production of proteins and other cellular components are a function both of the genes within the cells and of the environment in which the organism is developing and functioning.

The answer to the second question is that any one gene can exist in several forms that differ from one another, generally in small ways. These forms of a gene are called alleles. Allelic variation causes hereditary variation within a species. At the protein level, allelic variation becomes protein variation

DNA: It's a history book, a narrative of the journey of our species through time. It's a shop manual, with an incredibly detailed blueprint for building every human cell. And it's a transformative textbook of medicine, with insights that will give health care providers immense new powers to treat, prevent and cure disease. It is a molecule found in the nucleus of every cell. Every cell in our body whether lung cell or eye cell or toe cell, has same DNA. Sequence of DNA in any two people is 99.9% identical and only 0.1% is unique. It is made up of 4 subunits represented by the letters A, T, G, and C. The order of these subunits in the DNA strand holds a code of information for the cell. Just like the English alphabet makes up words using 26 letters, the genetic language uses 4 letters to spell out the instructions for how to make the proteins. Small segments of DNA are called genes. Each gene holds the instructions for how to produce a single protein. This can be compared to a recipe for making a food dish. A recipe is a set of instructions for making a single dish. An organism may have thousands of genes. The set of all genes in an organism is called a genome. A genome can be compared to a cookbook of recipes that makes an organism what it is. Proteins do the work in cells. They can regulate reactions that take place in the cell or can serve as enzymes, which speed-up reactions.

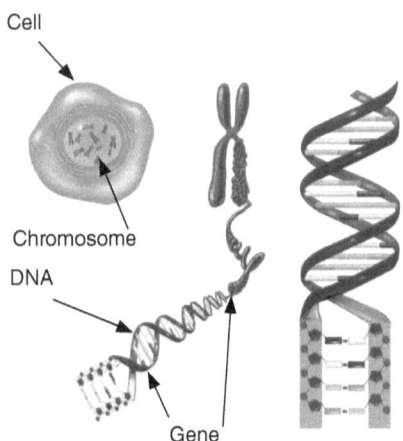

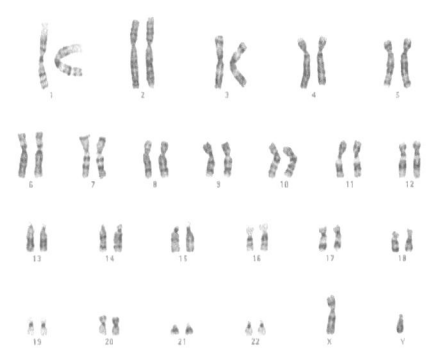

Humans have 23 pairs of chromosomes. Male DNA (pictured here) contains an X and a Y chromosome, whereas female DNA contains two X chromosomes.

Everything one sees in an organism is either made of proteins or the result of a protein action. Chromosomes are tiny packages of a cell and contain one DNA molecule and its associated proteins, called histones. Humans have 23 pairs of chromosomes, a total of 46. The molecule of DNA is made of two strands. Each strand is a chain of nucleotides which is formed of three components, a phosphate group (PO4-3), a pentose (5C sugar) and a nitrogen base (A, G, C, T). There are two strings of nucleotides coiled around one another in each chromosome and forms a double helix. C on one string is always opposite to G on the other string and A is always opposite to T. There are about 3.2 billion nucleotide pairs in all the human chromosomes. The order of the nucleotides carries genetic information, whose rules are defined by the genetic code, similar to how the order of letters on a page defines text information. Occasionally there is a kind of typographical error in a gene's DNA sequence. This mistake is called mutation which can cause a gene to encode a protein incorrectly or the error can mean no protein is made.

5.3 FUNCTION OF GENES

It is to provide information needed to make molecules called proteins in cells which are the smallest independent parts of an organism. Human body contains about 100 trillion cells. Genes give instructions and proteins carry out these instructions like building a new copy of a cell or repairing damage. A cell can make a copy of itself when cells divide. Each type of protein is a specialist that only does one job, so if a cell needs to do something new, it must make a new protein to do this job. Similarly, if a cell needs to do something faster or slower than before, it makes more or less of the protein responsible. Genes tells which protein to make and in what amounts. Proteins are made of a chain of 20 different types of amino acid molecules. This chain folds up into a compact shape, which is determined by the sequence of amino acids. It is this shape that determines what the protein does. The process of a DNA to become a protein follows a given processes.

    a. **Replication:** It is astounding to think that our bodies have trillions of cells while in the beginning it was just one cell. This massive expansion takes place by replication meaning making a copy of itself which is dividing into two new nearly identical cells. But before a cell divides itself in two new near identical cells, it must copy its DNA so that a complete set of gene passes onto each of the new cells. To make copy of itself, the helix unwinds and two strands separates. Now each strand becomes a model for making a new strand so the two new DNA molecules have one new and one old strand. The copy is done by a cellular protein machine called DNA Polymerase which reads the template strand and stitches a complimentary new strand. This process is known as Replication and is very fast and accurate. Occasional mistakes like deletion or duplication may occur and when not corrected, could lead to cancer or some genetic disorder.

b. **Transcription:** After copying, the next step is to make a protein. Transcription is the step when information stored into DNA is copied into RNA. A protein machine called RNA polymerase reads the DNA and makes the RNA copy. This RNA is called messenger RNA or mRNA. It may seem that same DNA is there in all the cells and hence all the body cells should also be similar. But on ground they are not the similar. For example a heart cell does not work in the same way as that of a skin cell. The reason is that RNA polymerase does not work alone but in conjunction with many helper proteins which vary in different body parts. That is why each cell does a different job as each cell turns on the appropriate gene for doing its necessary role.

c. **Translation:** Protein is not made directly by mRNA. Located in 23 pairs of chromosomes packed into the nucleus of a human cell, genes direct the production of proteins with the assistance of enzymes and messenger molecules. Specifically, an enzyme copies the information in a gene's DNA into a molecule called messenger ribonucleic acid (mRNA). The mRNA travels out of the nucleus and into the cell's cytoplasm, where the mRNA is read by a tiny molecular machine called a ribosome. Here the correct transfer RNA or tRNA is selected which has the correct amino acid which is building block of a protein. Proteins make up body structures like organs and tissue, as well as control chemical reactions and carry signals between cells. If a cell's DNA is mutated, an abnormal protein may be produced, which can disrupt the body's usual processes and lead to a disease such as cancer.

5.4 DNA AND RNA

When we think about heredity, DNA comes to mind first. It's true that DNA is the basic ingredient of our genes but RNA is the other form of genetic material inside our cells. Though both are types of genetic material, RNA and DNA are rather different. The chemical units of RNA are like those of DNA, except that RNA has the nucleotide uracil (U) instead of thymine (T). Unlike double stranded DNA, RNA usually comes as only a single strand. And the nucleotides in RNA contain ribose sugar molecules in place of deoxyribose. RNA is quite flexible, unlike DNA, which is a rigid, spiralstaircase molecule that is very stable. RNA can twist itself into a variety of complicated, three dimensional shapes. RNA is also unstable in that cells constantly break it down and must continually make it fresh, while DNA is not broken down often. RNA's instability lets cells change their patterns of protein synthesis very quickly in response to what's going on around them. Research has revealed that it is truly a remarkable molecule and a multitalented actor in heredity. RNAs can be grouped into two general classes. One class of RNAs is an intermediary in the process of decoding genes into polypeptide chains. We will refer to these "informational" RNAs as messenger RNAs because they pass information, like a messenger, from DNA to protein. Remaining minority of genes, the RNA itself is the final functional product. We will refer to these RNAs as "functional RNAs." As more is learned about the intimate details of cell biology, it has become apparent that functional RNAs fall into a variety of classes that play diverse roles. The main classes of functional RNAs contribute to various steps in the informational processing of DNA into protein. Two such classes of functional RNAs are found in prokaryotes and eukaryotes: transfer RNAs and ribosomal RNAs.

5.5 GENETIC VARIATION

If all members of a species have the same set of genes, how can there be genetic variation? The answer is that genes come in different forms called alleles. In a population, for any given gene there can be from one to many different alleles; however, because most organisms carry only one or two chromosome sets per cell, any individual organism can carry only one or two alleles per gene. The alleles of one gene will always be found in the same position along the chromosome. Allelic variation is the basis for hereditary variation. A useful classification of genetic variation could be *discontinuous* and *continuous* variation.

Discontinuous Variation: In this type of variation a character is found in a population in two or more distinct and separate forms called phenotypes. "Blue eyes" and "brown eyes" are phenotypes, as is "blood type A" or "blood type O." Such alternative phenotypes are often found to be encoded by the alleles of one gene. A good example is albinism in humans, which concerns phenotypes of the character of skin pigmentation. In most people, the cells of the skin can make a dark-brown or black pigment called melanin, the substance that gives our skin its colour ranging from tan colour in people of European ancestry to brown or black in those of tropical and sub- tropical ancestry. Although rare, albinos who completely lack pigment in their skin and hair, are found in all races. The difference between pigmented and unpigmented skin is caused by different alleles of a gene that encodes an enzyme involved in melanin synthesis.

Continuous Variation: It has an unbroken range of phenotypes in a population. Measurable characteristics such as height, weight, and skin or hair colour are good examples of such variation. Continuous variation is encountered more commonly than discontinuous variation in everyday life. We can identify examples of continuous variation, such as variation in size or shape, in plant, animal and in human populations. One area of genetics in which continuous variation is important is in plant and animal breeding. Many of the characters that are under selection in breeding programs, such as seed weight or milk production, arise from many gene differences interacting with environmental variation, and the phenotypes show continuous variation in populations.

5.6 GENES AND ENVIRONMENT

Genes cannot dictate the structure of an organism by themselves. The other crucial component in the formula is the environment. The environment influences gene action in many ways.

Genetic Determination Model: In this model, virtually all the differences between species are determined by the differences in their genomes. There is no environment in which a lion will give birth to a lamb. An acorn develops into an oak, whereas the spore of a moss develops into a moss. The two plants that result from these developmental processes resemble their parents and differ from each other, even though they have access to the same narrow range of materials from the environment. Within species, some variation is entirely a consequence of genetic differences that cannot be modified by any change in environment. The children from Africa brought to England will have dark skins, unchanged by the relocation of their parents.

Environmental Determination Model: Two identical twins, the products from a single fertilized egg produces two complete babies with identical genes. Let us say that the twins are born in Delhi but are separated at birth and taken to different countries. If one is reared in Singapore by Chinese-speaking foster parents, she will speak Chinese, whereas her sister reared in Moscow will speak Russian. Each will absorb the cultural values and customs of her environment. Although the twins begin life with identical genetic properties, the different cultural environments in which they live will produce differences between them and differences from their parents. Obviously, the differences in this case are due to the environment, and genetic effects are of no importance in determining the differences.

Genotype Environmental Interaction: In general, organisms differ in both genes and environment. To predict such a living organism, we need to know both the genetic constitution that it inherits from its parents and the historical sequence of environments to which it has been exposed. Every organism has a developmental history from conception to death. What an organism will become depends critically both on its present state and on the environment that it encounters during that moment and not only what environments it encounters, but also in what sequence it encounters them.

Genotype and Phenotype: In view of above models, one can now understand the terms genotype and phenotype. A typical organism resembles its parents more than it resembles unrelated individuals. It is often said "He gets his brains from his mother" or "She inherited diabetes from her father." Yet such statements are inaccurate. "His brains" and "her diabetes" develop through long sequences of events in the life histories of the affected people, and both genes and environment play roles in those sequences. In the biological sense, individuals inherit only the molecular structures of the fertilized eggs from which they develop. Individuals inherit their genes, not the end products of their individual developmental histories. To prevent such confusion between genes which are inherited and developmental outcomes, geneticists make the fundamental distinction between the genotype and the phenotype of an organism. Organisms have the same genotype in common if they have the same set of genes. Organisms have the same phenotype if they look or function alike. The genotype describes the complete set of genes inherited by an individual, while the phenotype describes all aspects of the individual's morphology, physiology, behaviour, and ecological relations. In this sense, no two individuals ever belong to the same phenotype, because there is always some difference, however small, between them in morphology or physiology.

5.7 GENOMICS

Most older geneticists began their careers trying to understand genes by working exclusively with their mutant phenotypes. They only dreamed that in their lifetimes they might see the hypothetical concept of a gene turned into clear reality, as both

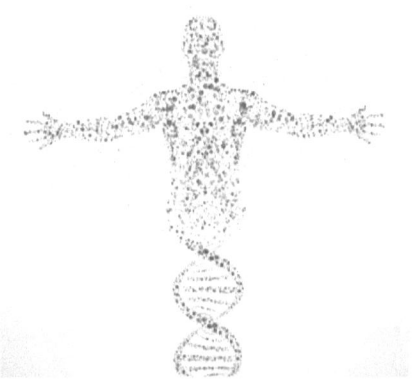

DNA sequence and function. The complete sequence of whole genomes was not even on their intellectual horizons. Yet today many genomes have been sequenced, with more on the way, and the use of these sequences has become routine in genetic analysis. Indeed the knowledge of entire genomes has revolutionized not only genetics but most fields of biological research. Now, we will examine the development and the operation of this exciting new field, broadly called genomics – the study of genomes in their entirety. The DNA sequence of the genome becomes the taking-off point for a whole new set of analyses aimed at the structure, function, and evolution of the genome and its components.

- *Bioinformatics* analyses the information content of entire genomes. This information includes the numbers and types of genes and gene products, as well as docking sites on DNA and RNA that allow functional products to be produced at the correct time and place.
- *Comparative genomics* considers the genomes of closely and distantly related species for evolutionary insight.
- *Functional genomics* uses various automated procedures to delineate networks of interacting genes active during some developmental process.

In humans, the genome is made up of around three billion nucleotide base pairs, which make up DNA molecules. Less than 2 per cent of the genome actually codes for proteins. The other 98 per cent is non-coding. Some of the noncoding sequences regulate the transcription of proteins and some are transcribed to RNA but do not get translated into protein. It is a staggering fact that humans only have about 20,000 protein-coding genes, the same as a starfish. Genomics is study of how the genes within the genome interact with each other and with the individual's environment.

Genetic testing is conducted when the researchers investigate a single piece of genetic information for specific bits of DNA with a known function. By investigating a single known entity, scientists may isolate the underlying causes of the specific genetic variant in question. Some examples of genetic or inherited disorders include cystic fibrosis, Down syndrome, hemophilia, Huntington's disease, phenylketonuria (PKU) and sickle-cell disease. In contrast to gene testing, Genomic testing is broader. It involves investigating large sections of genetic material and information, from which broad conclusions may be drawn. Some disorders and complex diseases that have been studied in the field of genomics include asthma, cancer, diabetes and heart disease. These diseases are caused by a combination of genetic and environmental factors, rather than simply a single genetic defect. The study of genomics has provided the medical community with new diagnostic tools and therapies for these complex diseases.

5.8 POPULATION GENETICS

Organisms do not live only as isolated individuals. They interact with one another in groups, populations, and there are questions about the genetic composition of those populations that cannot be answered only from a knowledge of the basic individual-level genetic processes. Why are the alleles of the protein Factor VIII and Factor IX genes that cause a failure of normal blood clotting, hemophilia, so rare in all human populations, whereas the allele of the hemoglobin gene that causes sickle-cell anemia is very common in some parts of Africa? What changes in the frequency of sickle-anemia are to be expected in the descendants of Africans in North America as a consequence of the change in environment and of the inter- breeding between Africans and Europeans and Native Americans? What genetic changes occur in a population of insects subject to insecticides generation after generation? What is the consequence of an increase or decrease in the rate of mating between close relatives? All are questions of what determines the genetic composition of populations and how that composition may be expected to change in time. These questions are the domain of population genetics. Population genetics is both an experimental and a theoretical science. On the experimental side, it provides descriptions of the actual patterns of genetic variation among individuals in populations and estimates the rates of the processes of mating, mutation, recombination, natural selection, and random variation in reproductive rates. On the theoretical side, it makes predictions of how the genetic composition of populations can be expected to change as a consequence of various forces operating on them.

5.9 QUANTITATIVE GENETICS

Ultimately, the goal of genetics is the analysis of the genotypes of organisms. But a genotype can be identified and therefore studied only through its effect on the phenotype. We recognize two genotypes as different from each other because the phenotypes of their carriers are different. Basic genetic experiments, then, depend on the existence of a simple relation between genotype and phenotype. That is why studies of DNA sequences are so important, because we can read off the genotype directly. In general, we hope to find a uniquely distinguishable phenotype for each genotype and only a simple genotype for each phenotype. At worst, when one allele is completely dominant, it may be necessary to perform a simple genetic cross to distinguish the heterozygote from the homozygote. Where possible, geneticists avoid studying genes that have only partial penetrance and incomplete expressivity because of the difficulty of making genetic inferences from such traits. However, most actual variation between organisms is quantitative, not qualitative. How do we study quantitative traits when they show such a complex relation between genotype and phenotype?

- Norm of reaction studies, in which different genotypes are allowed to develop in an array of different environments to determine the interaction of genotype and environment in the development of the character.
- Selection studies, in which successive generations are produced from the extreme individuals in the preceding generation.
- Heritability studies, in which the variation in the progeny of crosses is analyzed statistically to estimate the proportion of the variation in the original population that is a consequence of genetic differences and the proportion that is a consequence of environmental differences.
- Quantitative trait locus (QTL) studies, which associate phenotypic differences with alleles of a marker gene of known chromosomal location. Such an association with the marker gene reveals the approximate location of a gene affecting the quantitative character.

5.10 DNA SEQUENCING

The word sequencing refers to identification of the components of any story, such as the beginning, middle, and end, and also to recount the events in the order they occurred. The ability to sequence events in a text is a key comprehension strategy. Similarly DNA sequencing is the process of determining the precise order of nucleotides within a DNA molecule which is to determine the order of the four bases—adenine, guanine, cytosine, and thymine, in a strand of DNA. The sequence tells scientists the kind of genetic information that is carried in a particular DNA segment. Knowledge of DNA sequences has become indispensable for basic biological research, and in numerous applied fields such as medical diagnosis, biotechnology, forensic biology, virology and biological systematics. Researchers can use DNA sequencing to search for genetic variations and/or mutations that may play a role in the development or progression of a disease. The disease-causing change may be as small as the substitution, deletion, or addition of a single base pair or as large as a deletion of thousands of bases.

5.11 WHOLE GENOME SEQUENCING

WGS, full genome sequencing, complete genome sequencing, or entire genome sequencing is the process of determining the complete DNA sequence of an organism's genome at a single time. This entails sequencing all of an organism's chromosomal DNA as well as DNA contained in the mitochondria and in the chloroplast for plants. In the future of personalized medicine, whole genome sequence data will be an important tool to guide therapeutic intervention.

Genetic disorders: It is a disease caused in whole or in part by a change in the DNA sequence away from the normal sequence which can be caused by a mutation in one gene, mutations in multiple genes, combination of gene mutations and environmental factors, or by damage to chromosomes. We are learning that nearly all diseases have a genetic component. Some diseases are caused by mutations that are inherited from the parents and are present in an individual at birth, like sickle cell disease while other diseases are caused by acquired mutations in a gene or group of genes that occur during a person's life. Such mutations are not inherited from a parent, but occur either randomly or due to some environmental exposure such as cigarette smoke. Most genetic disorders are "multifactorial inheritance disorders," meaning they are caused by a combination of inherited mutations in multiple genes, often acting together with environmental factors. Examples of such diseases include many commonly-occurring diseases, such as heart disease and diabetes, which are present in many people in different populations around the world. Research on the human genome has shown that rare hereditary mutations in a single gene may cause or strongly predispose a person to these diseases run in a family and can significantly increase each family member's risk of developing the disease. One example is breast cancer, where inheritance of a mutated BRCA1 or BRCA2 gene confers significant risk of developing the disease.

5.12 IMPORTANCE OF GENETICS AND GENOMICS IN HEALTH

Both genetics and genomics play role in health and disease. Genetics helps individuals and families learn about how conditions such as sickle cell anemia and cystic fibrosis are inherited in families, what screening and testing options are available, and what treatments are available. Genomics research is helping to discover why some people get sick from certain infections, environmental factors, and behaviors, while others do not. For example, there are some people who exercise their whole lives, eat a healthy diet, have regular medical check ups, and die of a heart attack at age 40. There are also people who smoke, never exercise, eat unhealthy foods and live to be 100. Genomics may hold the key to understanding these differences. Apart from accidents such as falls, vehicle accidents or poisoning, genomic factors play a role in leading causes of death like heart disease, cancer, chronic lower respiratory diseases, stroke, alzheimer's, diabetes, nephritis. All human beings are 99.9 percent identical in their genetic makeup. Differences in the remaining 0.1 percent hold important clues about the causes of diseases. Gaining a better understanding of the interactions between genes and the environment by means of genomics is helping researchers find better ways to improve health and prevent disease. A person's health is influenced by his/her family history and shared environmental factors. This makes family history an important, personalized tool that can help identify many of the causative factors for conditions that also have a genetic component. My Family Health Portrait, https://www.genome.gov/17516481/, is a Web based tool from National Human Genome Research Institute, NHGRI, and the U.S. Surgeon General's Family History Initiative. The site helps create personal family health history. The case of famous actress Ms. Angelina Jolie will illustrate the importance of tracing one's family history. On February 16, 2013, at age 37, Jolie underwent a preventive double mastectomy after learning she had an 87% risk of developing breast cancer due to a defective BRCA1 gene. Her maternal family history warranted genetic testing for BRCA mutations as her mother, actress Marcheline Bertrand, had breast cancer and died from ovarian cancer. Her grandmother also died from ovarian cancer. The mastectomy lowered her chances of developing breast cancer to under 5 percent. Two years later, in March 2015, after annual test results indicated possible signs of early ovarian cancer, she underwent a preventive oophorectomy, as she had a 50% risk of developing ovarian cancer due to the same genetic anomaly. Jolie later wrote, "On a personal note, I do not feel any less of a woman. I feel empowered that I made a strong choice that in no way diminishes my femininity". Ms. Jolie's announcement of her mastectomy attracted widespread publicity and led to a global and long-lasting increase in

BRCA gene testing. Number of referrals has very significantly increased in many countries including India.

5.13 HUMAN GENOME PROJECT

The Human Genome Project, (https://www.genome.gov/12011238/) was the international, collaborative research program with a goal to completely map and understand all the genes of human being. All our genes together are known as genome which is unique for a given individual. Mapping the human genome involved sequencing a small number of individuals and then assembling these together to get a complete sequence for each chromosome. Therefore, the finished human genome is a mosaic, not representing any one individual. The $3billion project was formally founded in 1990 by the US Department of Energy and the

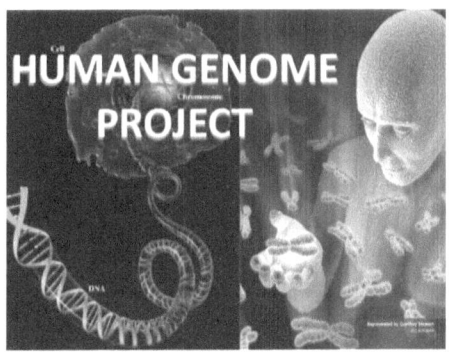

National Institutes of Health and was expected to take 15 years. Due to international cooperation and advances in the field of genomics as well as in computing technology, rough draft of the genome was finished in 2000. The Project was finally declared complete in April 2003. The sequencing of the human genome is benefiting diverse areas like molecular medicine, human evolution, understand diseases, genotyping of specific viruses, identification of mutations linked to different forms of cancer, design of medication, more accurate prediction of their effects, advancement in forensicsciences, biofuels and other energy applications, agriculture, animal husbandry, bioprocessing, risk assessment, bioarcheology and anthropology. A researcher can now investigate certain forms of cancer by narrowing down to a particular gene and then can examine what other scientists have written about this gene by visiting the human genome database on the

World Wide Webhttps://www.ncbi.nlm.nih.gov/genome/guide/human/. One major step toward such comprehensive understanding was the development of the HapMap (http://hapmap.ncbi.nlm.nih.gov/), which is a catalog of common genetic variation, or haplotypes, in the human genome. Another ambitious new initiative is to create The Cancer Genome Atlas (TCGA) (http://cancergenome.nih.gov/) in collaboration between the National Cancer Institute (NCI) and the National Human Genome Research Institute (NHGRI). It has generated comprehensive, multi-dimensional maps of the key genomic changes in 33 types of cancer. The TCGA dataset, comprising more than two petabytes of genomic data, has been made publically available, and this genomic information is helping the cancer research community to improve the prevention, diagnosis, and treatment of cancer.

5.14 TECHNOLOGY AND TECHNIQUES

Proteomics: Genomics, the study of whole genomes will soon make room for next big thing namely proteomics which is the study of all the proteins an organism makes. Protein does all the work in the body. They carry out all the functions that gene encode, so when a gene mutation occurs, the protein is what winds up being altered Given the link between genes and proteins, the study of proteins may end up telling more about the genes than the genes themselves. Proteomics can be used to reveal specific, abnormal proteins that lead to diseases, such as certain forms of cancer.

Pharmacogenetics and Pharmacogenomics: The terms "pharmacogenetics" and "pharmacogenomics" are often used interchangeably in describing the intersection of pharmacology (the study of drugs, or pharmaceuticals) and genetic variability in determining an individual's response to particular drugs. However there are differences. Pharmacogenetics is the field of study dealing with the variability of responses to medications due to variation in single genes by taking into account a person's genetic information regarding specific drug receptors and how drugs are transported and metabolized by the body. The goal of pharmacogenetics is to create an individualized drug therapy that allows for the best choice and dose of drugs. One example is the breast cancer drug trastuzumab (Herceptin). This therapy works only for women whose tumors have a particular genetic profile that leads to overproduction of a protein called HER2. Pharmacogenomics typically involves the search for variations in multiple genes that are associated with variability in drug response. This study may also examine genetic variation among large groups of people in order to see how different drugs might affect different racial or ethnic groups. Pharmacogenetic and pharmacogenomic studies are leading to drugs that can be tailor-made for individuals, and adapted to each person's particular genetic makeup. Although a person's environment, diet, age, lifestyle, and state of health can also influence that person's response to medicines, understanding an individual's genetic makeup is key to creating personalized drugs that work better and have fewer side effects than the one-size-fits-all drugs that are common today.

Stem Cell Therapy: Stem cell may hold the key to cuing brain and spinal cord injuries, part cure of cancer and many other medical problems. It is a hot research topic because of their totipotence, meaning that stem cells can turn into any kind of tissue. Stem cells are distinguished from other cell types by two important characteristics. First, they are unspecialized cells capable of renewing themselves through cell division. Second, under certain physiologic or experimental conditions, they can be induced to become tissue- or organ-specific cells with special functions. Embryonic stem cells (pluripotent stem cells) come from the embryo at a very early stage in embryo development (blastocyst). The stem cells in the blastocyst go on to develop all of the cells in the complete organism. Adult stem cells come from more fully developed tissues, like umbilical cord, blood in newly born, circulating blood, bone marrow or skin. Researchers are investigating the use of stem cells to repair or replace damaged body tissues, similar to whole organ transplants. Embryonic stem cells from the blastocyst have the ability to develop into every type of tissue (skin, liver, kidney, blood, etc.) found in an adult human. Adult stem cells are more limited in their potential (for example, stem cells from liver may only develop into more liver cells). In organ transplants, when tissues from a donor are placed into the body of a patient, there is the possibility that the patient's immune system may react and reject the donated tissue as "foreign." However, by using stem cells, there may be less risk of this immune rejection, and the therapy may be more successful.

Cloning: In the case of a cell, a clone refers to any genetically identical cell in a population that comes from a single, common ancestor. For example, when a single bacterial cell copies its DNA and divides thousands of times, all of the cells that are formed will contain the same DNA and will be clones of the common ancestor bacterial cell. Gene cloning involves manipulations to make multiple identical copies of a single gene from the same ancestor gene. Cloning an organism means making a genetically identical copy of all of the cells, tissues, and organs that make up the organism. There are two major types of cloning that may relate to humans or other animals: therapeutic cloning and reproductive cloning.

Therapeutic cloning involves growing cloned cells or tissues from an individual, such as new liver tissue for a patient with a liver disease. Such cloning attempts typically involve the use of stem cells. The nucleus will be taken from a patient's body cell, such as a liver cell, and inserted into an egg that has had its nucleus removed. This will ultimately produce a blastocyst whose stem cells could then be used to create new tissue that is genetically identical to that of the patient. Reproductive cloning is a related process used to generate an entire animal that has the same nuclear DNA as another currently or previously existing animal. The first cloned animals were frogs. Dolly, the famous sheep, is another example of cloning. The success rates of reproductive animal cloning, however, have been very low.

5.15 BIOINFORMATICS

We live in the information age with practically everything available at the fingertip. But for genetics, it is information overflow. The greatest challenge facing the molecular biologist is to make sense of the wealth of data that has been produced by the genome sequencing projects. Gene sequencing, generation, subsequent storage, interpretation and analysis are entirely computer dependent tasks. Huge increase in the scale of data of the genomic era has necessitated incorporation of computers into this process. The first challenge is the intelligent and efficient storage of this mass of data followed by reliable access to this data. It is essential that these databases are easily accessible and that an intuitive query system is provided to allow researchers to obtain very specific information on a particular biological subject. The data should be provided in a clear, consistent manner with some visualisation tools to aid biological interpretation. Today any one can access the massive biological databases at www.ncbi.nlm.nih.gov to reach the National Centre for Biotechnology Information and search the result of entire Human Genome Project.

Cancer informatics   Gene regulation
Personalized medicine   Protein modeling
Computational biology
Image analysis
Comparative genomics   Gene expression databases
Epidemic models   Computational drug discovery

Bioinformatics

Sequence analysis   Bio-ontologies and semantics
Evolution and phylogenetics   Structure prediction
Cheminformatics   Next generation sequencing
Computational intelligence   Transcriptomics
Biomedical engineering Amino acid sequencing
Structural bioinformatics   Medical informatics
Microarrays
Visualization

The data itself is meaningless before analysis and the sheer volume present makes it impossible for even a trained biologist to begin to interpret it manually. Therefore, incisive computer tools must be developed to allow the extraction of meaningful biological information. Bioinformatics gives ready access to powerful analytical tools. Gene sequencing, generation, subsequent storage, interpretation and analysis are entirely computer dependent tasks. Scientists also need to be able to integrate the information obtained from the underlying heterogeneous databases in a sensible manner in order to be able to get a clear overview of their biological subject. Once all of the biological data is stored consistently and is available to the scientific community, the requirement is then to provide methods for extracting meaningful information from the mass of data. Here comes Bioinformatics, which uses software programs as tools to lay bare the huge genomic data in a scientific way to the research community.

Major categories of Bioinformatics Tools: There are both standard and customized products to meet the requirements of particular projects. Data-mining software retrieves data from genomic sequence databases and visualization tools analyze and retrieve information from proteomic databases. These can be classified as homology and similarity tools, protein functional analysis tools, sequence analysis tools and miscellaneous tools. Few of these bioinformatics is done with sequence search programs like BLAST, sequence analysis programs, like the EMBOSS and Staden packages, structure prediction programs like THREADER or molecular imaging/modelling programs like RasMol and WHATIF. MOOC platforms provide online certifications in bioinformatics and related disciplines, including Coursera's Bioinformatics Specialization (UC San Diego) and Genomic Data Science Specialization (Johns Hopkins) as well as EdX's Data Analysis for Life Sciences XSeries (Harvard).

A relatively new field, translational bioinformatics, is poised to become an important discipline for precision medicine. With the advent of the X-ray, and later of magnetic resonance and other imaging technologies, enabled visualization of tissues and organs has become possible. Each of these technological advances necessitates a companion advance in the methods and tools used to analyze and interpret the results. Defined by the American Medical Informatics Association as "the development of storage, analytic, and interpretive methods to optimize the transformation of increasingly voluminous biomedical data, and genomic data, into proactive, predictive, preventive, and participatory health," translational bioinformatics is at the forefront of data-driven health care.

Advances in biological methods and technologies have opened up a new realm of possible observations. The invention of the microscope enabled doctors and researchers to make observations at the cellular level. Advent of advanced X-ray, magnetic resonance and other imaging technologies, has enabled visualization of tissues and organs possible. Each of these technological advances necessitated a companion advance in the methods and tools used to analyze and interpret the results. With the increasingly common use of technologies like DNA and RNA sequencing, DNA microarrays, and high-throughput proteomics and metabolomics, comes the need for novel methods to turn these new types of data into new information and that new information into new knowledge. That new knowledge, in turn, gives rise to action, providing insights regarding how to treat disease and ideally how to prevent it in the first place. TBI can be categorised into four major areas such as clinical "big data", or the use of electronic health record (EHR) data for discovery (genomic and otherwise); genomics and pharmacogenomics in routine clinical care; other omics for drug discovery and repurposing; and personal genomic testing, including a number of ethical, legal, and social issues that arise from such services. In summary, we are entering a new era in data-driven health care. Translational bioinformatics methods will make a difference in patients' lives. The infrastructure, information technology, policy, and culture need to catch up with the technological advances. Working at the cutting edge of translational bioinformatics will open opportunities abound and a bright future.

5.16 GENETIC ENGINEERING

Thanks to recombinant DNA technology, genes can be isolated in a test tube and characterized as specific nucleotide sequences. But this achievement is not the end of the story. Actually knowledge of a sequence is often the beginning of a fresh round of genetic manipulation. When characterized, a sequence can be manipulated to alter an organism's genotype. The introduction of an altered gene into an organism has become a central aspect of basic genetic research, but it also finds wide commercial application. One example of the latter is plants kept from freezing by the incorporation of arctic fish "antifreeze" genes into their genomes. The use of recombinant DNA techniques to alter an organism's genotype and phenotype in this way is termed genetic engineering. It refers to any changes in genetic makeup that result from the direct manipulation of DNA using various technical methods. The process of genetic engineering is intended to produce a useful or desirable characteristic in an organism and on a molecular level and may include additions, deletions, or targeted changes to the genome. More simply put, genetic engineering involves cutting, pasting, and/or editing DNA to produce a valuable effect. Interestingly, these alterations can involve introduction of genetic material from either the same or from different type of organism. Dr. Paul Berg and colleagues are credited with creating the first ever recombinant DNA molecule in 1972. For this work, Dr. Berg was awarded the 1980 Nobel Prize in chemistry for "his fundamental studies of the biochemistry of nucleic acids, with particular regard to recombinant-DNA". Several other terms are also used to describe the technology, like gene manipulation, gene cloning, recombinant DNA technology, genetic modification, and the new genetics.

Though there are many diverse and complex techniques involved, the basic principles of genetic manipulation are reasonably simple. The premise of the technology is based on the fact that genetic information, encoded by DNA and arranged in the form of genes, is a resource that can be manipulated in various ways to achieve certain goals in both pure, applied science and medicine. There are many areas in which genetic manipulation is of value like basic research on gene structure and function, production of useful proteins by novel methods, generation of transgenic plants and animals, medical diagnosis and treatment, genome analysis by DNA sequencing.

Genetic engineering, as explained above, works by physically removing a gene from one organism and inserting it into another, giving it the ability to express the trait encoded by that gene. It is like taking a single recipe out of a cookbook and placing it into another cookbook. The process broadly works as follows:

1. First identify an organism that contain the desired traits.
2. DNA is extracted from that organism. This is like taking out the entire cook book.
3. Now the desired gene has to be located out of the entire DNA which was extracted. This is known as Gene Cloning.
4. The gene may be modified slightly to work in a more desirable way when put inside the recipient organism.
5. The new gene(s), called a transgene is delivered into cells of the recipient organism. This is called transformation. The most common transformation technique uses a bacteria that naturally genetically engineer plants with its own DNA. The transgene is inserted into the bacteria, which then delivers it into cells of the organism being engineered. Another technique, called the gene gun method, shoots microscopic gold particles coated with copies of the transgene into cells of the recipient organism.
6. Once a transgenic organism has been created, traditional breeding is used to improve the characteristics of the final product. So genetic engineering does not eliminate the need for traditional breeding. It is simply a way to add new traits to the pool.

Few examples of genetic engineering are as follows:

- Until the 1980s the only source of insulin available to diabetics was from animals slaughtered for meat and other purposes. The supply was never large enough to provide a sufficient amount of affordable insulin for everyone who needed insulin. In 1982, however, the U.S. Food and Drug Administration approved insulin produced by genetically altered organisms.
- Some of the products produced by rDNA techniques are human growth hormone (for children whose growth is insufficient because of genetic problems), alpha interferon (for the treatment of diseases), interleukin-2 (for the treatment of cancer), factor VIII (needed by hemophiliacs for blood clotting), erythropoietin (for the treatment of anemia), tumor necrosis factor (for the treatment of tumors), and tissue plasminogen activator (used to dissolve blood clots).
- Genetic engineering also promises a revolution in agriculture. Recombinant DNA techniques enable scientists to produce plants that are resistant to herbicides and freezing temperatures, that will take longer to ripen, and that will manufacture a resistance to pests, among other characteristics.
- The science and art of animal breeding also are likely to be revolutionized by genetic engineering. For example, scientists have discovered that a gene in domestic cows is responsible for the production of milk. Genetic engineering makes it possible to extract that gene from cows producing large volumes of milk or to manufacture that gene in the laboratory. The gene can then be inserted into other cows whose milk production may increase by dramatic amounts because of the presence of the new gene.
- One of the most exciting potential applications of genetic engineering involves the treatment of human genetic disorders. Medical scientists know of about 3,000 disorders that arise because of errors in an individual's DNA. Conditions such as sickle-cell anemia, Tay-Sachs disease, Duchenne muscular dystrophy, Huntington's chorea, cystic fibrosis, and Lesch-Nyhan syndrome result from the loss, mistaken insertion, or change of a single nitrogen base in a DNA molecule. Genetic engineering enables scientists to provide individuals lacking a particular gene with correct copies of that gene. If and when the correct gene begins

functioning, the genetic disorder may be cured. This procedure is known as human gene therapy (HGT).

- In 1990, a research team at the National Institutes of Health (NIH) attempted HGT on a four-year-old SCID patient. The patient received about one billion cells containing a genetically engineered copy of the gene that his body lacked. Another instance of HGT was a procedure, approved in 1993 by NIH, to introduce normal genes into the airways of cystic fibrosis patients. By the end of the 1990s, according to the NIH, more than 390 gene therapy studies had been initiated. These studies involved more than 4,000 people and more than a dozen medical conditions.
- In 2000, doctors in France claimed they had used HGT to treat three babies who suffered from SCID. Just ten months after being treated, the babies exhibited normal immune systems.
- One of the most controversial uses of genetic engineering has been cloning or producing a genetically identical copy of an organism. While the ethics of cloning are hotly debated, the first ever sheep named Dolly was cloned in 1996 by scientists.
- Glow-in-the-dark cats – It sounds strange, but in 2007, scientists in South Korea altered the DNA of a kitty so that its fur would glow in the dark, and then cloned other cats from it, making the world's first glowing cats.
- Plants that fight pollution – Poplar trees developed by scientists at the University of Washington can absorb polluted water through their roots and clean it before the water is released back into the air. The plants were many times more efficient at cleaning certain pollutants than regular poplars.
- Golden rice – Genetic modification is often used to make "healthier" foods, such as golden rice, which contains beta-carotene – the very same vitamin that makes carrots orange. The result is that people without access to many vitamins will get a healthy dose of vitamin A when the rice is consumed.
- Faster-growing trees – Demand for wood can be met by trees that grow faster than average. Genetic engineering has produced trees that can ward off biological attacks, grow more quickly and strongly, and create better wood than trees that are not genetically modified.

> Bigger, longer-lasting tomatoes – When tomatoes are genetically engineered, they can be made bigger and more robust. These are engineered to produce tomatoes that can remain fresh for longer, can be shipped farther from where they are grown, and can be harvested all at the same time rather than harvesting only parts of a field at each harvest.

Quite a few institutions in India are doing quality research in the area of genetic engineering like IISc Bangalore, TIFR Mumbai, Biotechnology departments of IITs, ICGEB New Delhi, IGE Kolkata, IMT Chandigarh, NCCS Pune.

However controversy remains and human gene therapy is the source of great controversy among scientists and non scientists alike. Few individuals maintain that the HGT should not be used. If we could wipe out sickle cell anemia, most agree, we should certainly make the effort. But HGT raises other concerns. If scientists can cure genetic disorders, they can also design individuals in accordance with the cultural and intellectual fashions of the day. Will humans know when to say "enough" to the changes that can be made with HGT?

5.17 SYNTHETIC BIOLOGY

The term "synthetic biology" can be considered as a basket of techniques that represent an engineering approach to design and construct novel biological parts, devices and systems, and to redesign existing biological systems and organisms. In contrast to conventional genetic engineering which involves mixing and matching pieces of DNA from different organisms, synthetic biology aims to use synthetic (man-made) DNA, "rewritten" genetic sequences to create new forms of life or to reprogram existing organisms to produce chemicals or perform other tasks that they would not otherwise do. Few outcome of synthetic biology, are yeast cells

equipped with synthetic gene sequences that cause the organism to secrete chemical compounds including vanillin, stevia or saffron flavorings; reengineered microbes that produce biofuels; and a synthetic self-replicating bacterium, whose creators claimed it to have been the first completely artificial life form. Scientists working in synthetic biology believe that it could dramatically change both the society and the natural world in which we live. George Church, Professor of Health Sciences and Technology at Harvard University and MIT notes that "people talk themselves out of things very easily. Things that they think are a million years away or never, are actually four years away." Synthetic biologists are experimenting with a variety of rapidly developing approaches, often in combination:

Bio Bricks: They are small, interchangeable biological units, standardized sequences of DNA with a particular function that can be put together, like toy LEGO® building blocks. These bricks are used to design and assemble synthetic biological circuits, which can then be incorporated into living cells to construct new biological systems.

Xeno DNA: All natural DNA have four "bases," but researchers have invented artificial DNA with six bases, which may have applications in research and medicine. Xeno DNA has been shown to survive in a living cell.

Digital DNA and Gene Synthesis: Much of the work of constructing or changing genetic sequences is done on computers, and then converted to physical output as artificial DNA. There is now an industry devoted to supplying customized genes on demand, and it is becoming increasingly capable of synthesizing viruses and eventually more complex organisms. Synthetic biologists speak of designing life from scratch. So far, this entails "mimicking" life by generating artificial DNA and then inserting it into an existing cell to replace some of the genetic material previously there.

3D Biological Printing: What if we could create physical organisms that have been digitally designed? 3D printing is made possible by fusing layers upon layers of materials made from durable plastics and metals based on a template, designed with a 3D Computer Aided Design (CAD) software. With this technology, and a 3D printer, one can create designs or print 3D models of just about anything under the sun, provided templates are available. Experimental 3D printing of cancerous cells was reported in April 2014. Venter and others assume that remote 3D printing of even more complex biological structures will eventually become standard practice.

 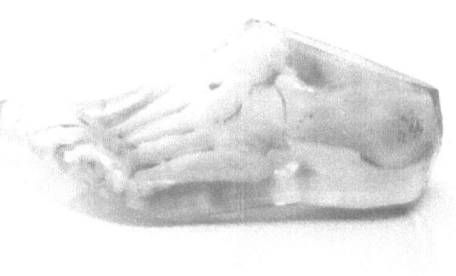

3D scan of an unborn child is taking on a whole new meaning. Instead of a picture of ultrasound, a Japanese company is now giving a 'Shape of an Angel', a 3D print of your foetus. The 3D model is created by 3D processed image data with Bio Texture. With 3D printing, doctors would have a cheaper alternative to learn about the human anatomy and also be able to inject realism into surgical practices without the use of cadavers. Since the printing of these medical models is so accurate, surgeons can also plan a surgery on a printed model like this before the real patient goes under the knife.

Gene Editing: For years, scientists have been able to delete very short segments of DNA in an organism, or turn off specific genes. To insert new DNA, viruses are used as transporters into the cell and the genome. But this method is highly inaccurate. Today, several new and more sophisticated techniques are being developed to delete or replace specific pieces of DNA, both in petri dishes and in organisms. At present, the biologically engineered system known as CRISPR/Cas9 is generating the most interest of several gene editing techniques.

Gene Drives: One emerging application of synthetic biology and gene editing is known as a "gene drive." This system would permit the genetic alteration of populations of organisms, and thus of entire ecosystems. It relies on the introduction of "selfish genetic elements that can increase the odds that they will be inherited," as noted by Harvard's Kevin Esvelt, combined with other genetic changes that might be thought desirable, for instance to block malaria transmission by mosquitoes. Normally, the introduced gene would have a 50% chance of being passed on to the next generation, but a gene drive increases these odds dramatically.

Human Applications: Human applications, particularly in healthcare, are among the most prominent goals of synthetic biologists, investors and regulatory agencies. A headline from a publication of the Biotechnology Industry Organization takes a typically optimistic view "Synthetic Biology is like The Sword in the Stone to Defeat Devastating Diseases". These applications are focused on curing or preventing disease.

Medical Diagnostics: Human applications, particularly in healthcare, are among the most prominent goals of synthetic biologists, investors and regulatory agencies. Scientists at Stanford University and the University of Montpellier, France, reported in 2015 that they had modified bacteria to recognize glucose levels in urine and change color, thus potentially providing a test for diabetes. So far, it is seen as a proof of principle. Synthetic biology is creating a diabetes detection system that works internally as can be seen from the optimism of Stanford's Prof. Drew Endy, "Why not make our medicines from biology directly? We foresee that global health can be practically and affordably realized using biology. If you need more medicine somewhere, people can simply grow it where and when it is needed."

Vaccine Production: Synthetic biologists are currently testing how to rapidly produce vaccines. The vision, envisages mobile to sequence strain samples from around the world and automatically upload that data into a database, allowing scientists anywhere to access the information and create or re-create vaccines without needing the physical strain sample.

Xenotransplantation: This refers to the use of organs from other animal species to replace faulty ones in humans. Efforts are being made to genetically modify pigs, and possibly other animals, to create organs that are at least partly "humanized" and thus available for research or even for clinical use. Among others, Craig Venter's Synthetic Genomics is working on this project with Lung Biotechnology Inc., a subsidiary of United Therapeutics. Venter optimistically stated, "We are re-engineering the pig, changing its genetic code. If we succeed with rewriting the pig genome, we will have replacement organs for those who need them".

Synthetic biology in India is generally confined to few institutions and groups whereas large number of institutes and groups are working in life sciences and biotechnology. Centre for Systems and Synthetic Biology, based in University of Kerala, Trivandrum is one such specialized organisation. Another special interest Group on Synthetic Biology in India, Sanjeevani is based at Jawaharlal Nehru University, New Delhi. There are individual scientists and groups working on different aspects of synthetic biology in some institutes including the Indian Institutes of Technology (IITs), IISc, National Centre for Biological Sciences (NCBS), and Council of Scientific and Industrial Research (CSIR) laboratories.

Synthetic biology pioneers have ambitious visions for the techniques they are developing like "rebuild the living world." Major technical problems remain and reality may fall short of such vision. New genetic technologies are powerful and complex. Much of the basic science of genetics is still far from well understood. There is therefore a need to consider multitude of ethical, moral and scientific issues to ensure that they won't do more harm than good.

5.18 EPIGENETICS

One of the biggest challenges to Mendel's law comes from a phenomenon called *epigenetics*. The prefix epi means over or above. In epigenetics, organism with identical alleles may exhibit different phenotypes. The difference in phenotypes doesn't come from the genes themselves but from elsewhere in the chemical structure of the DNA. The tiny chemical tag, called methyl groups, attached to DNA acts like operating system and tells the programme how often to work, where and when. In case of epigenetics, the tag can shut a gene down or turn a gene on. Some epigenetics effects are normal and beneficial. But there are other tags which act like mutations and cause disease. Many traits like eye color, body shape and even some personality quirks are written in our genes as inherited from the parents. But the story does not end there. Many environmental factors like, where we live, what we eat, how much we exercise can affect our gene expression. Epigenetics is the study of these heritable alterations that are not due to changes in DNA sequence or DNA code, a change in phenotype without a change in genotype, which in turn affects how cells read the genes. In simple terms, certain circumstances in life can cause genes to be silenced or expressed over time. In other words, they can be turned off (becoming dormant) or turned on (becoming active). Epigenetics makes us unique. With 20,000+ genes, possible permutation-combination of the resultant arrangement of genes being turned on or off is enormous! Today, a wide variety of health and behavioral issues, have some evidence linking them with epigenetic mechanisms, including cancers of almost all types, cognitive dysfunction, and respiratory, cardiovascular, reproductive, autoimmune, and neurobehavioral illnesses. Known or suspected drivers behind epigenetic processes could be many agents, including heavy metals, pesticides, diesel exhaust, tobacco smoke, polycyclic aromatic hydrocarbons, hormones, radioactivity, viruses, bacteria, and basic nutrients. However if we can map every single cause and effect of the different combinations, and can reverse the gene's state to keep the good while eliminating the bad, then we could theoretically cure cancer, slow aging, stop obesity, and much more.

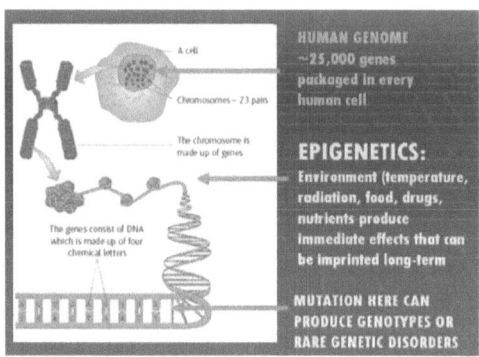

An analogy to understand epigenetics goes like this. Let us say that human life span is a very long movie. The cells would be the actors and actresses, essential units that make up the movie. DNA, in turn, would be the script, instructions for all the participants of the movie to perform their roles. Subsequently, the DNA sequence would be the words on the script, and certain blocks of these words that instruct key actions or events to take place would be the genes. The concept of genetics would be like screenwriting. The concept of epigenetics, then, would be like directing. The script can be the same, but the director can choose to eliminate certain scenes or dialogue, altering the movie considerably. Just like, Steven Spielberg's finished product would be drastically different than Woody Allen's for the same movie script!

5.19 PERSONALISED MEDICINE AND DIGITAL PATIENT

One of the biggest cause of death is adverse reaction to medication which is supposed to help patients. This unnecessary death is explained in terms of pharmacogenomics, the analysis of human genome and heredity to determine how drug work in individual people. The good news is that personalised medicine will be a new brand of care that is being designed to fit the unique genetic makeup of each individual patient. The flip side is that nobody knows how many genes are involved in disease and how many genes can cause the same disease. Epigenetics further complicates the matter by turning genes off and on in unexpected way. However, we are nearing the biggest revolution in drug delivery. It is going to meet two most important things. First it will meet the requirement of predicting and detecting disease well before it becomes life threatening and second the medicine will work for us individually for our unique body. Adverse drug reactions occur as most common drugs go everywhere in the body. Only a small proportion ends up at a disease site where it can have beneficial effect. Other 99% can cause problems in previously healthy tissues. A 17-year-old Kolkata boy was prescribed Dapsone, an antibiotic, for a skin problem, and developed severe, potentially fatal, side effects. It took visits to two speciality hospitals, a fortnight's stay in an intensive-care unit, and a week in a ward to bring the boy back from the brink of death (www.indiaspend.com). Within India, the ADR (adverse reporting rate) reporting rate has more than doubled in the last few years to 40 in 2015 but it is lower than 130, the average ADR reporting rate for high-income countries (www.ncbi.nlm.nih.gov/pmc). India addresses the problem of adverse drug reactions by ignoring or not reporting the data. But the problem do exists. More than 50% of Americans take at least one prescription drug each day and as a direct result, fourth highest cause of death in North America is due to bad reaction to a drug resulting in more than 100,000 deaths per year. In addition to the potential bad reaction, many of these drugs simply do not work for the individual who takes them. All these problems spring from a one size fits all approach to medicine. The doctor does not know details of patient's genetic code and proteomics and therefore cannot knowhow the patient will respond to the drug and state of the immune system. In short, the doctor does not have access to lot of molecular level information of the patient which may lead to wrong diagnoses and inappropriate intervention. So the patient and doctor embark on a trial and error journey. Knowledge of oneself at the molecular level is going to have huge impact on us. When we were born, we were given a unique body which is possibly the most sophisticated organism on the planet but we were not given the operator's manual or warning lights we have in a car. So we don't get a warning light about onset of a life threatening disease in the body or an immune system going off the track or an environmental exposure which will lead to arthritis in the future.

Moleculer based personalized medicine will give the desired operator's manual. It will contain four categories of information viz first is sequence of genome which contains the blue print of the individuals physical being. Second, proteomes and about 100 or more proteins in the blood should give a good snapshot of health. Third will be metabolome, 10 or more metabolites of the blood will tell as to how the body is dealing with diets and diagnose disease. Fourth is microbiome or bacteria and other micro-organism that lives in and on the body. Detection of few hundred bacteria in the feces will provide diagnostic and therapeutic information, particularly about immune disorder.

Convergence of multiple technologies is taking place and that is making possible personalized medicine. It is various branches of biology, computer science and technology and its many variants, internet, cloud, big data, remote sensing and nanotechnology. It is really all encompassing digital revolution. Creating digital patient means converting his/her molecular information into numbers. For example, once genome is digitized, it will be able to answer whether the body will have reaction to a drug prescribed by doctor or if the drug will be effective to a specific body. Digital patient can also find out about risk of inherited disease, a cancer patient will be able to get the cancer cell genome sequenced, find out mutating genes and decide on a drug for treatment. Social media and patient to patient will permit comments from other patients with similar cancer genetic profile about efficacy of their treatment. This will mean that the patient will start asking important questions about his/her health and monitor the trend. For example, a patient will be able to see the trend of proteome of the blood. An increase in inflammatory proteins or retinol binding protein will give warning about onset of arthritis or diabetes. Eventually such biomarker proteins will provide early warning of everything from heart failure, kidney failure, lung infection, stroke, cancer and all. Out of balance metabolome and microbiome may indicate overuse of antibiotic, inheritance of unhealthy type of bacteria from parent or wrong diet. These data will be complemented with other data which will be gathered from wearable. Once all the information is collected, it needs to be stored and analysed. Storing genome of roughly 3 billion base pairs require about 800 MB of storage. Genes are often sequenced many times which will mean data of 100 GB for a single human genome. It will be costly to store. But one can store just the difference with a common reference genome which will reduce the cost substantially. Now comes one of the most promising area of Bioinformatics as it pertains to interpreting large database of genomic, proteomic and other 'omic' information. So digital patient can get a pretty detailed molecular level information and store it digitally either in cloud or in personal hard drive. Now comes big data. A digital patient will be able to compare with the data of thousands of other digital patients. Comparison of genotype (individual's genome) with phenotype (individual's personal traits) over millions of individuals will reveal how various genes contribute to every aspect from colour of hair to tendency to lisp to athletic potential. Correlation between genotype with particular disease and their environment will bring to light subtle relationship between environment and individual susceptibility to disease. "Dr. Google" is being hugely researched about health-related issues. Virtual communities like PatientsLikeMe (www.patientslikeme.com) launched its first online community for ALS patients in 2006. From there, the company began adding other communities for other life-changing conditions, including multiple sclerosis, Parkinson's disease, fibromyalgia, HIV, chronic fatigue syndrome, mood disorders, epilepsy, organ transplantation, progressive supranuclear palsy, multiple system atrophy, and Devic's disease. In April 2011, the company expanded its scope and opened its doors to any patient with any condition. In 2017, PatientsLikeMe has entered into partnership with iCarbonX to apply next generation biological measures and machine learning to understand more about the basis of human health and disease. Today the website covers more than 3000 health conditions, with new members joining daily from the US and other countries around the world. Sites like Patientslikeme, CureTogether are providing a forum for patients to share their stories as to how well their treatments are going on. It will help to identify best therapies as patients start sharing their digital selves, become a virtual clinical trial, research new drugs. The first manifestation of digital patient is electronic medical record (EMR) or electronic health record (EHR). Health Record India (www.healthrecordsindia.com) is a cloud based solution that collects, digitises, optimises any type of medical records (Prescriptions, Xrays, ECG, CT Scans etc.) and stores them in a secure and easy-to-access way. My Health Records (www.myhealthrecords.in) is a health management services for safely storing, viewing and sharing of health related records and information online. The site is now connected with Microsoft® HealthVault® and migrating accounts to HealthVault.

Company specialises in digitization of medical records and provides complete automated services for patient's records management. SMS alerts/reminders can be set for multiple events. They have a growing panel of reputed doctors and experts for online consultations. Michael Snyder of Stanford University learned from his genetic analysis that he was at risk of developing type 2 diabetes though he had no such family history (https://psmag.com/social-justice/making-it-personal-geneticist-michael-snyder-puts-a-face-to-personalized-medicine-57554). He got his molecular-self studied and also did proteomic and metabolomic analysis of blood over a period of time. He interpreted the data. In response, he altered the diet and increased exercise routine. After six months his glucose level came back to normal without any drug treatment. Thus he proactively managed his own health and has shown a pointer as to how medicine of the future will be managed. Future doctors will not only physically examine a patient but also the digital part of the patient. Doctors will interpret the data, point out any wrong trend and suggest ways to reverse it. So, the role of the doctor will also be that of a health coach/mentor. The identification of mutations in the genome is enabling the development of individualized therapies. We have already described the case of actress Angelina Jolie. Analysis of her digital self and subsequent treatment had dropped the chances of breast cancer from 87 percent to 5 percent. She told her children that they need not fear of losing her due to breast cancer.

Future of Personalised Medicine: Personalised medicine and gene therapy go hand in hand. Gene therapy says that if we can detect the genetic basis of a disease, that disease can be treated by inserting a new gene in place of defective one. However this idea is not a simple one as our body has elaborate defence mechanism to prevent injection of any invading DNA or RNA. Research is going on in this area. An exciting development is nanosurgery of DNA to correct defects by so called CRISPR (clustered regularly interspersed short palindromic repeats) technologies. This field is limited to lack of full understanding of how brain works and relating the biology of brain to behaviour. There are approximately 11 billion of neurons in the brain having an average 7000 connections with other neurons. These neurons are connected through synapses which transmit electrical signals from one neuron to next. So there could be 100 trillion individual synapses that can be firing at any time. It is a formidable task to map the firing of hundreds of thousands of synapses. New technologies are coming up with innovative solutions like nano-sized sensors which can be implanted deep inside the brain. Information processing demands are also huge. Imaging the activity of all neurons in a mouse brain could generate 300 petabyte of data in an hour. Data generation, storage and analysis of human brain will require considerably more. Research is going on and future will see that new imaging techniques will map brain activity with our moods, behavior and actions. This will, in turn, lead to very personalized therapies for depression, addiction, schizophrenia and host of other disorders. Next is personalized medicine to tackle aging. Research is aimed at understanding and treating aging process. Hayflick Limit, named after Leonard Hayflick of Stanford University, is a phenomenon of number of times a normal human cell will divide until cell division stops. Evidence shows that the telomeres associated with each cell's DNA will get slightly shorter with each new cell division until they shorten to a critical length and this aging of the cell population appears to correlate with the overall physical aging of the human body. Children suffering from progeria undergo premature aging and die, essentially of old age, in early teens. These children have a mutation in a gene that results in rapid shortening of telomeres which is growing area of research. 2009 Nobel Prize in Medicine was awarded jointly to geneticist Elizabeth H. Blackburn, Carol W. Greider and Jack W. Szostak for identification of a protein called telomerase that can lengthen the telomere to let the cells keep on dividing. Research is on to find a drug that activates telomerase. Small molecule activators have shown to improve the apparent health status of mice. Meditation, exercise and diet also play a direct role in activating telomerase. A potential downside of telomerase activation is that large number of tumour cell also exhibit telomerase activity and can divide indefinitely, thus increasing cancer risk. However it is a matter of time to fully understand the biology at the molecular level and using this understanding to reengineer the tissues as they age. Anti-aging is drawing huge attention and funding. A company called Human Longevity Inc (www.humanlongevity.com) has a goal to make a 100 year old to look like 60. Google has entered the area by a company called Calico (https://www.calico) with the goal of combating aging and associated disease

Leroy Hood (https://en.wikipedia.org/wiki/Leroy_Hood) one of biology's living legends, has pioneered systems biology, and his vision is "scientific wellness," which aims to improve health and prevent disease by combining personalized behavior coaching with DNA and blood testing, activity tracking, and other measures. He and colleagues have compiled "personal, dense, dynamic data clouds" for 108 people after tracking them for 9 months including their full genome sequences; blood, saliva, urine, and stool samples taken every 3 months that measured levels of 643 metabolites and 262 proteins; and physical activity and sleep monitoring. The massive data set could help people avoid diabetes and other health issues. Indeed, Hood intends to move forward with his previously proposed 100K Wellness Project, for which he hopes to recruit 100,000 people for by 2020. Hood hopes to develop predictive models to delineate early biomarkers for disease and early intervention. Focus is on prevention rather than treating a disease.

DIY Gene Testing Kits: New drug treatments based on genomic research are likely to take time. However, screening and diagnostic tests are becoming available, including DIY kits that can be bought online by members of the public. A home-based gene testing kit called 23andMe is available online at a reasonable price. One has to register and fill in some information on the company's website (23andme.com) and spit into a tube that is sent to their laboratories for analysis. Xcode Life Sciences (http://xcode.in) is another personalized genomics company in India and offers genetic tests for life style diseases. They also give genetic counselling and nutrition and fitness recommendations based on genetic risk and current life style pattern.

Genomic medicine is going to become more mainstream and doctors need to know what tests are available. This need to educate healthcare professionals in genomic medicine has been highlighted by Professor Ian Cumming, chief executive of Health Education England "The genomic revolution is critical to the personalisation of healthcare, to the prediction of disease and to the prevention of disease. That's why we will make sure we train not just our current workforce but our future workforce in the skills they need to be able to respond." The future is genomics and, like it or not, we are part of it.

5.20 INTERNET OF THINGS IN HEALTH

Internet: ARPANET adopted TCP/IP on January 1, 1983, and from there researchers began to assemble the "network of networks" that became the modern Internet. The online world then took on a more recognizable form in 1990, when computer scientist Tim Berners-Lee invented the World Wide. On August 15, 1995, Videsh Sanchar Nigam Limited (VSNL) launched public Internet access in India. In 2016, there were three billion internet user globally, almost half the world's population. Social network reach about 80% of users in developing and developed economies alike.

Exponential Growth is an immensely powerful concept. An ancient Indian chess legend beautifully explains the concept. The legend goes that the tradition of serving Paal Paysam to visiting pilgrims started after a game of chess between the local king and the lord Krishna himself. The king was a big chess enthusiast and had the habit of challenging wise visitors to a game of chess. One day a travelling sage was challenged by the king. To motivate his opponent the king offered any reward that the sage could name. The sage modestly asked just for a few grains of rice in the following manner: the king was to put a single grain of rice on the first chess square and double it on every consequent one. Having lost the game and being a man of his word the king ordered a bag of rice to be brought to the chess board. Then he started placing rice grains according to the arrangement: 1 grain on the first square, 2 on the second, 4 on the third, 8 on the fourth and so on. Following the exponential growth of the rice payment the king quickly realized that he was unable to fulfil his promise because on the twentieth square the king would have had to put 1,000,000 grains of rice. On the fortieth square the king would have had to put 1,000,000,000 grains of rice. And, finally on the sixty fourth square the king would have had to put more than 18,000,000,000,000,000,000 grains of rice which is equal to about 210 billion tons and is allegedly sufficient to cover the whole territory of India with a meter thick layer of rice. At ten grains of rice per square inch, the above amount requires rice fields covering twice the surface area of the Earth, oceans included. It was at that point that the lord Krishna revealed his true identity to the king and told him that he doesn't have to pay the debt immediately but can do so over time and to this day visiting pilgrims are still feasting on Paal Paysam and the king's debt to lord Krishna is still being repaid. Internet has reached into the second half of the chess board. As the rice pile starts to rival Mount Everest magnitude when it reaches 64th square, rapidly evolving internet will both enrich and overwhelm. Business needed to make a choice by rising to the challenge of a new internet driven environment. Internet is seen as the biggest technological disruption of all time after invention of printing press in 1450. Next major technological revolution is Internet of Things (IoT).

Internet of Things: Actual term "Internet of Things" was coined by Kevin Ashton in 1999 during his work at Procter & Gamble in supply chain optimization. He wanted the company to link RFID information to the internet. The concept was simple but powerful. If all objects in daily life were equipped with identifiers and wireless connectivity, these objects could communicate with each other and be managed by computers. Because the internet was the hottest new trend in 1999, he called his presentation "Internet of Things". At the time, this vision required major technology improvements. After all, how would we connect everything on the planet? What type of wireless communications could be built into devices? What changes would need to be made to the existing Internet infrastructure to support billions of new devices communicating? What would power these devices? What must be developed to make the solutions cost effective? There were more questions than answers to the IoT concepts in 1999. Today, many of these obstacles have been solved. Cisco's Internet of Things Group predicts there will be over 50 billion connected devices by 2020. IoT describes a system where items in the physical world, and sensors within or attached to these items, are connected to the Internet via wireless and wired Internet connections. These sensors can use various types of local area connections such as RFID, NFC, Wi-Fi, Bluetooth, and Zigbee. Sensors can also have wide area connectivity such as GSM, GPRS, 3G, and LTE. According to the McKinsey report "The Internet of Things is such a sweeping concept that it is a challenge to even imagine all the possible ways in which it will affect business, economies, and society." Cisco has expanded the definition of IoT to the Internet of Everything (IoE), which includes people, places, objects and things. Internet of Things will:

> Connect both inanimate and living things: Today IoT connects everything from industrial equipment to everyday objects. It can also include living organisms such as plants, farm animals and people. Wearable computing and digital health devices, such as Nike+ Fuel band and Fitbit, are examples of how people are connecting in the Internet of Things landscape.

> Use sensors for data collection: The physical objects that are being connected will have one or more sensors which will monitor a specific condition such as location, vibration, motion, temperature. Wearable electronic devices, with their built in sensors, worn on the head, neck,

arms, torso, and feet will connect to each other and to systems that can understand and present information.
- ➤ IoT Hardware: Servers, desktop, tablet, and cellphone remain integral parts as the command center and remotes. Other key connected devices are standard network devices like routers and switches.
- ➤ IoT Software: It addresses key areas of networking and action through platforms, embedded systems, partner systems, and middleware. They are responsible for data collection, device integration, real-time analytics, and application and process extension within the IoT network.

Internet of Things Applications: Internet of things promises many applications in human life, making life easier, safe and smart. Few such applications are given below.
- ➤ Smart Cities: Many major cities were supported by smart projects, like Seoul, New York, Tokyo, Shanghai, Singapore, Amsterdam, Dubai, Bhubaneswar, Pune, Jaipur, Surat, Ludhiana, Kochi, Ahmedabad, New Delhi. These cities are careful planned in every stage, with support of governments, citizens and other agencies to implement the internet of things technology in every aspect like infrastructure, public transportation, reduced traffic congestion, citizen safety, improved health and more engaged community.
- ➤ Smart Home and Buildings: Wi-Fi have started becoming part of the home IP network and due the increasing rate of adoption of mobile computing devices like smart phones, tablets, etc. Many companies are considering developing platforms that integrate the building automation with entertainment, healthcare monitoring, energy monitoring and wireless sensor monitoring in the home and building environments. Some interesting application of IoT in smart homes and buildings are smart lighting, smart environmental and media, air control and central heating, energy management and security.
- ➤ Smart Energy and the Smart Grid: A smart grid is related to the information and control for smart energy management and will permit real time two way communication between suppliers and consumers, creating more dynamic interaction on energy flow, which will help deliver electricity

more efficiently and sustainably. Many applications can be handling due to the internet of things for smart grids, such as industrial, solar power, nuclear power, vehicles, hospitals and cities power control.

- Smart Health: Smart health sensors are used to collect comprehensive physiological information, use cloud for storage and then send the data wirelessly to caregivers for further analysis and review. It replaces the process of having a health professional come by at regular intervals to check the patient's vital signs, instead provides a continuous automated flow of information. In this way, it simultaneously improves the quality of care through constant attention and lowers the cost of care by reduces the cost of traditional ways of care in addition to data collection and analysis. Many people around the worlds are suffering from the bad health because they don't have ready access to effective health monitoring and may be a suspected to be as critical situation patients. But with small, powerful wireless solutions connected through the IoT is now making possible for monitoring to come to these patients.

- Smart Transportation and Mobility: Road condition monitoring and alert application is one of the most important of IoT application. The main idea is to apply the principles of crowd sourcing and participatory sensing… The smart transportation is deal with three main conceptions transportation analytic, transportation control, and vehicle connectivity. The transportation analytic represents the analysis of demand prediction and anomaly detection. The routing of vehicles and speed control in addition to traffic management are all known as transportation control. IoT can also be used in transportation as an electric vehicles,

- Smart Factory and Smart Manufacturing: Smart factory added a new value in manufacturing revolution by integrates artificial intelligence, machine learning, and automation of knowledge work. These advances in the way machines and other objects communicate and the resulting way in which decision-making moves from humans to technical systems means that manufacturing becomes smarter.

- Smart Environment: Environment plays a major role in the life of human, animals, birds, fishes and plants. They get adversely affected in unhealthy environment which needs smart ways and new technologies for monitoring and management. Smart environment is an important technology in our everyday life which provides many facilities and solutions for many environmental applications such as water and air pollution, weather and radiation monitoring, waste management, natural disaster, and many other environment indicators which could get connected to each person through home area network. Applications of internet of things in environment can broadly be divided to two main categories like environmental resources management, and environmental quality and protection management. IoT can provide a high resolution, and accuracy for weather monitoring by data exchange and information sharing. It can avoid or reduce the impact of a large number of natural disasters that affect in many aspects of life through the distribution of a number of sensor systems for many types of natural disasters and linking these systems with disaster management system. IoT will provides a means of smart agriculture and add great potential in resource saving. By using sensors networks, and scientific research databases, growing of plants and other agriculture products can monitored by managing weather, water and sunlight. In addition, the IoT for environmental monitoring can aid in measuring emissions from factories and detect forest fires.

IoT in Health Care: When we need to measure our health parameters, we go to lab and provide blood sample or to a clinic where they measure blood pressure, ECG, radiology imaging and perform other diagnostic tests. We may have to wait long for these tests to happen due to long queue. After that we bring the result to the doctor. We thus rely on health care institutions to obtain data about our own body. What if such parameters are available right away? This idea has driven the development of technologies, gadgets, sensors that can be embedded, implanted or worn in order to measure health parameters. Sensors can be embedded in tissue (pacemaker), ingestible (smart pills), epidermal (smart skin), wearable (clothing, jewelry), and external (smart watches). Professor Takao Someya of University of Tokyo, is developing novel organic bio devices which will be able electrical and chemical signals of billions of neurons and will give real time high resolution visualization of neural activities. Future belongs to digestible, embedded and wearable sensors. These sensors will measure vital signs and important health parameters 24×7, transmitting data to the cloud and sending alerts to medical system. Patients of gastrointestinal disease can swallow sensor that include a video camera which could render an instant diagnosis by combining lab result with colonoscopy pictures. Equivital (www.equivital.co.uk) has developed a sensor that continuously monitors vital signs like pulse oxemetry, oxygen saturation and core temperature which are risk factors for adverse events of dialysis. MC10 (www.mc10inc.com) uses cutting edge technology to create the most intelligent, flexible platform for biometric healthcare analytics and is partnering with the University of Rochester to unite the company's powerful technological capabilities in physiological sensing and pattern recognition algorithms with the university's clinical expertise and commitment to big data analytics. Toray, a Japanese company is developing a cloth containing nanofibres coated in a transmittable layer. Hexoskin (www.hexoskin.com) has developed electronic shirt that tracks movement, respiration and pulse. Professionals like astronauts, athletes can see their body metrics displayed on a smartphone. Such statistics will generate detailed recommendation on training and general health. The smart bra has successfully been tested and shows 74% correlation to actual cancer stages in all types of breast tissues. National Taiwan University has developed a sensor which can be embedded in a tooth cavity and measure jaw movement, drinking, chewing, coughing, speaking and even smoking. Results are wirelessly transmitted and can potentially curb assorted addiction like smoking and drinking. Leslie Saxon from University of Southern California's Keck School of Medicine says that a baby born ten years from now could be tattooed with an integrated circuits at birth that could monitor ECG, nutritional status, physical activities, sleep, breathing rate, body temperature and hydration that can be used in health management and disease prediction. Qualcom Tricorder X Prize (www.tricorder.xprize.com) was announced in 2012 promising to award $10 million for making a device that should diagnose 15 different medical conditions ranging from sore throat to colon cancer across 30 people in 3 days. Final Frontier Medical Devices and Dynamical Biomarkers Group were announced winners of the 2017 Qualcomm Tricorder XPRIZE. Cloud DX was recognized as XPRIZE's first Bold Epic Innovator receiving $100,000, sponsored by Qualcomm Foundation. The potential medical use of smartphone application and devices can be described by a real life story of Dr. Eric Topol, Chief Academic Officer at Scrips Health, was on a flight from Washington DC to San Diego when pilot whther there was a doctor on board as a passenger is having severe chest pain. Dr. Topol put his iPhone into AliveCor ECG bracket, performed an ECG and concluded that patient was having a heart attack. An emergency landing somewhere near Cincinnati followed and the patient survived after a stent implant. It is said that AliveCor has been used more than a dozen times in-flight to make diagnoses from myocordial infraction to atrial fibrillation.

These are just a few examples of IoT-based healthcare solutions. Many more are emerging. But as one reporter has noted, "The real vision for the future is that these various smaller applications will converge to form a whole. Imagine that you are a relative of a patient who forgot medicine. You receive an alert, locate their location, remotely check their vital signs remotely to see if they are seriously ill, navigation system of the car locates hospitals which have free beds, clearest traffic route to get there and finally even a parking slot."

Enabling Technologies:
- Smart sensors: This combines a sensor and a microcontroller, making possible to harness accurate measurement, monitoring and analyzing a variety of health status indicators. These such as heart rate, blood pressure, glucose level and oxygen saturation in the blood. Smart sensors can even be incorporated into pill bottles and connected to the network to indicate whether a patient has taken a scheduled dose of medication. For smart sensors to work effectively, the microcontroller components must incorporate several essential capabilities like low-power operation for small device footprint, integrated precision-analog capabilities for high accuracy, and graphical user interfaces for improved usability.
- Gateways: They are the information hubs that collect sensor data, analyze it and then communicate it to the cloud via wide area network technologies. Gateways can be designed for clinical or home settings; in the latter, they may be part of larger connectivity resource that also manages energy, entertainment and other systems in the home. Medical device designers can also use the platform to create remote-access devices for remote monitoring.

- Wireless networking: It removes the physical limitations on networking imposed by traditional wired solutions. Microcontrollers support wireless connectivity for devices based on popular wireless standards such as Bluetooth and Bluetooth Low Energy (BLE) for personal area networks (PAN) and Wi-Fi® and Bluetooth for local area networks (LAN) in clinics or hospitals.

5.21 FEW GENETICS COURSES, SOCIETIES, JOURNALS
Courses
- **Harvard University** www.harvard.edu
  Graduate programme in Biological and Biomedical Scienc with various specialisation such as genetics/Genomics/Bioinformatics, Microbiology, Molecular Biology
- **Massachusetts Institute of Technology** www.mit.edu
- **United States Cambridge**, MA
  Graduate programme in Biological Scienc with various specialisation such as genetics/Genomics/Bioinformatics, Molecular Biology
- **Johns Hopkins University** www.jhu.edu
  Graduate programme in Biological Science with specialisation in Cell Biology, Molecular Biology
- **Stanford University** http://med.stanford.edu/genetic-counseling.html
  MS Program in Human Genetics and Genetic Counseling. It is a two years programme.
- **Queen's University, UK** www.qub.ac.uk
  M.Sc in Bioinformatics & Computational Genomics. It is one year course.
- **University of Maryland** www.medschool.umaryland.edu/genetics/
  Masters in Genetic Counseling, MS in Human Genetics, Master's in Cellular and Molecular Biomedical Science
- **Indian Institute of Science Bangaluru** http://www.iisc.ernet.in/ug/biology.html
  B.Sc in Biology by research, 4 years
  The Bachelor of Science (Research) Programme in biology at the Institute offers a unique blend of Biology courses with the courses in Chemistry, Physics, Math, Engineering, and Humanities and prepares its graduates to take up challenging issues in research in Biological Sciences.

- **IIT Bombay** http://www.bio.iitb.ac.in/
  M.Sc in Biotechnology duly supported by the Dept. of Biotechnology, Govt. of India
- **IIT Kharagpur** http://www.iitkgp.ac.in/department/BT
  Offers B Tech & M Tech in Biotechnology and Biochemical Engineering
- **IIT Delhi** http://beb.iitd.ac.in/academics.html
  B Tech & M Tech in Biotechnology and Biochemical Engineering & M.Sc by rsearch and ofcours Ph D
- **Institute of Genetic Engineering, Kolkata,** http://www.ige-india.com/
  B.Sc and M.Sc courses in Medical Biotechnology and Human Genetics
- **Banaras Hindu University** www.bhu.ac.in/science/
  M.Sc in Molecular & Human Genetics
- **Punjab University** www.pu.edu.pk/program/
  BS & MS in Microbiology & Molecular Genetics

Genetic Societies

- Human Genomics in Global Health, WHO http://www.who.int/genomics/policy/regulatory_bodies/en/
- Genetic Society of America http://www.genetics-gsa.org/
- British Society of Genetic Medicine http://www.bsgm.org.uk/
- European Society of Human Gentics https://www.eshg.org/home.0.html
- Human Genetic Society of Australisia https://www.hgsa.org.au/
- Indian Society of Human Genetics http://www.ishg.co.in/
- Genetic Society of Japan http://gsj3.jp/eng/e-msg.html
- National Human Genome Research Institute https://www.genome.gov/10002335/regulation-of-genetic-tests/
- Advisory Committee on Genetic Testing, UK http://www.oecd.org/sti/biotech
- Board of Gnetic Counselling, India http://www.geneticcounselingboardindia.com

Genetic Journals

- Nature Review Genetics:*www.nature.com/nrg*
- Nature Genetics:*www.nature.com/ng*

- Genome Research: *genome.cshlp.org/*
- Genes & Development: *genesdev.cshlp.org/*
- PLOS Biology: *journals.plos.org/plosbiology*
- Trends in Genetic: *www.journals.elsevier.com/trends-in-genetics*
- Genome Biology: *//genomebiology.biomedcentral.com/*
- Bioinformatics: *https://academic.oup.com/bioinformatics*
- Bioinformatics and Computational Biology: *www.worldscientific.com* ›
- *Genetic Engineering & Biotechnology:* www.journals.elsevier.com/journal-of-genetic-engineering-and-biotechnology

5.22 CONCLUSION

In the past few hundred years we have discovered the laws of gravity, electricity, theory of general relativity, telephone, vaccinations, and many more. All these discoveries were external to us. Only recently in the 20th century we have started delving into the building blocks of what we are made of. Today we are on the verge of changing the course of humanity through the ability to manipulate our genetic building blocks; the technology about ourselves. Thomas Hunt Morgan proved Mendel's recessive and dominant trait through extensive studies of fruit fly. In 1915, he went on to combine his and Mendel's findings in The Mechanism of Mendelian Heredity, which is considered the foundation of modern genetics. Boom of genetics came in 1977 when British biochemist Fred Sanger and his team sequenced DNA for the first time in history using the Sanger Method. Dolly the sheep became the first animal ever cloned in 1996. In 2003, Human Genome project was completed, where scientists successfully mapped 92% of the human genome at 99% accuracy. Now, Google will even sequence and store our genome for a mere few dollars a year. Genetically Modified Organism, or GMO, is the modern application of genetics to plants and animals. Over 50% of plants in the US are GMO plants. By removing and inserting experimental genes, crops have developed resistance to pests/herbicides, increased nutrition, higher yields, and even production of new drugs. Bill Gates has donated $15 million for the development of a vitamin packed, bio-fortified GMO banana, which is supposed to be grown in poverty stricken areas of Africa and India to combat nutritional deficiencies.

Super Humans: Taking all that we've learned from our mistakes, our successes, and exponential technology growth, we are at a point in time where genetically engineered enhancements to humans is becoming a reality. With increased understanding comes technological progression to the point when scientists can physically manipulate DNA with Nano tools, and thus our genetic code which will accrue two key benefits like decreasing negative genetic mutations and increasing positive genetic mutations. Some people are naturally born with positive random genetic mutations that, after rigorous research, could potentially be inserted into the population. The other area of genetic engineering that could save lives is decreasing negative genetic mutations of genetic diseases like Cystic Fibrosis, Breast/Colon Cancer, Huntington's Disease, and many more. Identifying the gene responsible for these diseases early in life, scientist could presumably cut out or destroy these mutated genes that can cause terrible problems later in life The possibilities are endless. China has taken a lead in this field. The Chinese BGI Cognitive Genomics Project is sequencing DNA from 1,000 high IQ individuals around the world in order to be able to predict genes with high correlation with intelligence. They could then allow couples to select their "most-intelligent eggs" and effectively increase IQ's from 5-15 points per generation.

The possibility of creating superior intelligence is exciting as well as scary. Instead of furthering humanity, genetics could also be the next devastating weapon. The countries with the most advanced genetic engineering would have the strongest, fastest, and smartest soldiers as well as general population. Endless possibilities always go hand in hand with endless dangers if not regulated and used responsibly. Only time will tell how much we will be able to unravel the greatest mysteries of ourselves.

REFERENCES

1. Genes, A Short Introduction, Jonathan Slack
2. An Introduction to Genetic Engineering, Desmond Nichol
3. Molecular Biology of the Gene, James D Watson, Tania A Baker
4. The Age of Genomes, Steven Monroe Lipkin
5. Wiki Genetics https://en.wikipedia.org/wiki/Genetics
6. Wiki Human Genetics https://en.wikipedia.org/wiki/Human_genetics
7. Wiki Geneticist https://en.wikipedia.org/wiki/Geneticist
8. Wiki Genetic https://en.wikipedia.org/wiki/Genetic
9. Wiki Advanced Genetics ag.teamdna.de/wiki/mc17/index.htm

10. SingularityHub https://singularityhub.com/topics/#sm.00007zgjojzetde7piu1f48cxulmj
11. About Nanotechnology https://www.foresight.org/nano/
12. Nature Genetics https://www.nature.com/ng/
13. Nature Journal https://www.nature.com › nature
14. High Impact Journal for genetics and genomics massgenomics.org/2013/01/high-impact-journals-genetics-and-genomics.html
15. Futurism https://www.futurism.com

EPILOGUE

I feel an urge to cross the finish line by penning a few lines about the necessity of the book for both the young and the not so young – 'it is all about disruption'. We have been trained to think of the future as a linear extension, typically imagining change as a 10 percent improvement from where we are today. But the pace of change is accelerating fast and potential disruptive technologies are looming in the horizon. For example, machine learning and big data are reaching a critical momentum. Advances in speech and image recognition of recent time completely dwarf what came before. Even as we begin to embrace such changes in a particular technology, far more exciting possibilities are unfolding by convergence of multiple technologies. In such circumstances, the future will be a very broad set of possibilities and these will unfold surprisingly quickly. We need to develop a strong understanding of technology, survey new innovations, forecast their pace, gauge implications and adopt strategy and tools to change. Basically, we need to have a mindset of a *futurist as well as that of a technologist.*

Nurturing curiosity is the first step. To know how technology will disrupt career, employment, industry and society, we need to explore what is in the pipe line, read outside the comfort zone, attend innovation camps and learn through reverse mentoring. Whatever it be, the goal is to develop a healthy obsession with technology. One of the challenging aspects of a futurist is tolerance for ambiguity, to get comfortable with the reality that we cannot predict the future hundred percent.

Microsoft's CEO Satya Nadella believes in installing a culture of discovery for the success of the company. He says, "we want to push to be more of a learn-it-all culture than a know-it-all culture".

Exponential leaders, young and not so young, use the skills of a futurist to improve lives of the people they touch and society as a whole. It means being sensitive by investing in humane policies and being positive.

When Google's high altitude balloons connect the most rural and underdeveloped areas to universal high speed internet and drones deliver medical supplies after natural disasters, we start believing how technology can give wings to our imagination.

Technology is and will increasingly be a driver of disruption, create new opportunities, destabilize few existing ones and create not yet imagined values. We need to be vigilant about potential sources of disruption. The answers may not be perfectly clear but it is important to seek relentlessly.

Happy reading.